MARCHER
WALKER
PILGRIM

A MEMOIR FROM THE GREAT MARCH FOR CLIMATE ACTION

12/14/18
For a sustainable future for all life,
Ed Fallon

Marcher, Walker, Pilgrim is published by Business Publications Corporation Inc., an Iowa corporation.

Copyright 2018 by Ed Fallon.

Reproduction or other use, in whole or in part, of the contents without permission of the publisher and author is strictly prohibited.

ISBN-978-0-9986528-6-3
Business Publications Corporation Inc., Des Moines, IA

Business Publications Corporation Inc.
The Depot at Fourth
100 4th Street
Des Moines, IA 50309
515-288-3336

All proceeds from the sale of *Marcher, Walker, Pilgrim* go to support the work of Climate March, 501(c)(3) successor to the Great March for Climate Action.

"This book is intended to serve as my historical recounting of the Great March for Climate Action. The narrative references my own notes from the March and includes the stories and memories of other participants or those we met along the way. The book also reflects present-day recollections of past experiences in my life that were personally influential or are relevant to the story of the March.

"Every effort has been made to provide an enjoyable, truthful, although certainly incomplete, telling of the story of the March. In a few cases, courtesy and readability have made it necessary to change someone's name or alter minor details of the story.

"While the information has been compiled carefully to ensure maximum accuracy at the time of publication, it is provided for general guidance only and is subject to change. The publisher cannot guarantee the accuracy of all information or be responsible for omissions or errors.

"So, if something in this book pisses you off, don't blame the publisher. You'll have to sue the author's impoverished ass — and good luck with that."

— *Ed Fallon*

MARCHER
WALKER
PILGRIM

A MEMOIR FROM THE GREAT MARCH FOR CLIMATE ACTION

ED FALLON

DEDICATION

To the marchers, staff, and supporters of the Great March for Climate Action

TABLE OF CONTENTS

INTRODUCTION
1. Failure
2. Why Not a March!
3. Second-Class Travel
4. Post-Victorian Man
5. Base Camp
6. In the Shadow of Valero
7. The Streets of LA
8. Wind and Sand
9. Moonwalk
10. The Honeymoon is Over
11. Sleep
12. Payson
13. Food
14. Hypothermia
15. Zuni Enchantment
16. One Earth Village
17. Walking Stick
18. Albuquerque
19. Farmer Miller
20. Marcher
21. Father and Son
22. Roadkill
23. Storms on the Prairie
24. Independence
25. Crossing Keystone
26. Across the Water
27. "Don't Be Stupid"
28. Grace
29. Son and Father
30. My Piano
31. Home
32. The Great River
33. Walker
34. Sacrifice Zone
35. Longest Meets Largest
36. Collision
37. Resilience
38. My Tent
39. Puzzle
40. Forgiveness and Redemption
41. Coal Country
42. I Leave the March
43. Pilgrim
44. Broken
45. The White House

INTRODUCTION

"Formerly, only a few men wrote valuable books. Now, anybody writes and prints anything he likes and poisons people's minds."

— M.K. GANDHI, *HIND SWARAJ*

Humanity has traveled a long way since the printing press. Today's writers are no more gifted than writers of the past, yet there are certainly a lot more of us. If Gandhi fretted over the access the printing press gave to the masses, we can only imagine his distress had he known of the Internet.

When I set out to write a book in 2009, one modest goal was to not "poison people's minds." A more serious goal, and perhaps one that would have pleased Gandhi, was to share my story about the rewards and challenges of a life of public service.

The flow of that literary focus was redirected in 2014 by the Great March for Climate Action, an idea that sprang from my experience with the Great Peace March in 1986. I had been eager to join the Peace March but, due to a series of back injuries, I couldn't even walk half a mile let alone across the country. So I signed on as an organizer, and a notepad, landline, and old manual typewriter became my life for the next six months.

Despite its near collapse early on, the Peace March was a success. It infused new life into the nuclear test ban movement, spawned citizen-led initiatives in US-Soviet diplomacy, and inspired thousands of Americans to deepen their commitment to peace. The experience transformed the marchers and, for me, solidified a life-long commitment to public service.

Similarly, I knew the Climate March would change the lives of its participants. I knew it would have an impact on my life as well. It did, but in ways I never could have imagined.

Walking across America is an enormous challenge. Pounding pavement, gravel, and sand 15 to 20 miles a day for eight months is unkind to one's body, even under the best conditions. The difficulties are markedly greater when one walks on a schedule, as we did, with rallies and events planned along the route weeks and even months in advance.

When you walk on a schedule you can't take time off for an illness. You can't rest for a week to recover from an injury. You can't take a leave of absence for a funeral or a wedding. You simply keep going or you quit.

It became important for me to try to walk every step of the way. Sure, there's something gratifying about traversing the breadth of a continent on foot. But far more important, I saw the March as a moving dramatization of the urgency of the climate crisis. Our steps were sacrificial performance art, showcasing the deep individual commitment we all must make to assure our collective survival in the New Climate Era.

So, this book had no choice but to evolve. It's become the story of my experience on the Great March for Climate Action, and how the March changed my life — and in a small way, may have helped change history.

With the Climate March as the conduit, this narrative evolved in another more deeply personal way as well. It became the story of my search for love and meaning. That search took unexpected twists during eight months of walking, interacting with people and places in ways not available to one traveling by car, train, or even bike.

My search for love is, perhaps surprisingly, a relatively new quest — one that began in earnest after my second divorce in 2010. Since then, with every subsequent failed relationship, I learned something new and important, gaining experience I hoped would prevent me from botching the next opportunity to find deep, lasting love.

My search for meaning, on the other hand, goes back to my pre-teen years, when such a quest was disparaged simply as odd behavior. I recall someone asking my mother, "When did Eddie start acting weird?" Without batting an eye, she responded, "Age 11."

True enough. At 11, the weirdness began. I gave up soda pop and hotdogs. More significantly, I became almost mystically devout — praying fervently, embracing the symbols and rituals of the Catholic faith to a degree that was, well, weird for anyone, let alone an eleven-year-old boy.

Then, just as dramatically as it started, this short-lived mystical encounter crumbled. Doubt filled the void. Skepticism eroded the facade of faith, and with an intoxicating blend of recklessness and exhilaration, I abandoned any pretense that Catholicism had even an iota of meaning in my life.

I sealed my exit from the Catholic Church in dramatic fashion during Mass at age sixteen while seated in the balcony with my brother Bill. The parish priest, Father Comaine, openly despised and ridiculed young people. Every sermon provided a new opportunity for him to rage against us. That day's tirade was particularly vitriolic and had me furious. When he bellowed at the congregation, "Young people today think God is somebody who rides on a motorcycle," my right hand, seemingly of its own volition, thrust its middle finger toward the spiteful priest, no doubt securing me a balcony seat in the great fiery beyond.

During the Great March for Climate Action, as I walked what seemed an impossible distance, the quest for love and meaning intertwined with my daily struggles against the raw power of nature, with quarrels with other marchers, and with our efforts to rouse a lethargic America to mobilize against the changing climate that threatens our very survival.

"Ok," you say. "That's great that you want to share this incredible adventure of walking across the United States with us and remind us of the urgency of climate change. But really, why burden us with tales of juvenile navel-gazing and your pathetic love life?"

Like wine, writing is truth serum, possessing the power to impart unanticipated clarity and understanding to one's experiences. It liberates even as it leads to discomfort. It exposes fears and aspirations one would rather leave undisturbed. Writing facilitates self-discovery, and I've often found myself unprepared to embrace the self that writing discovers.

So, perhaps I write this book for selfish reasons as well: to find answers to my quest for love and meaning.

I have one final motive in writing this book, and that is to reflect on the importance of home — both this planetary home we share with all life and the more intimate, personal "home-place" we share with our family, coworkers, and closest friends.

On the Climate March, we traveled through astounding natural beauty. We met deeply generous people who welcomed and embraced us nearly everywhere we went. Yet I felt a strong, continual longing to be home. And home for me is two places — Iowa and Ireland. As a child and young man, Ireland sowed the seeds of my love for the land. Ireland is my spiritual refuge, the place which, in my youth, connected me to my historical past, fired my imagination, and taught me the value of both farming and community.

Iowa has been my physical home since 1984. I've traveled enough to know that few places on Earth are as blessed as this lush, verdant crescent between the Missouri and Mississippi rivers. Iowa's subtle, fertile beauty is a treasure beyond comprehension to the masses of agriculturists around the world, many who toil in soil that barely deserves the name. Iowa is a garden that warrants tending, preservation and, in far too many cases, restoration.

While every city and hamlet on the planet are home to friendly, compassionate people, nowhere in my experience compares with Iowa and Ireland, where farming is not only economically important but socially respected. And Iowans, like the Irish, are a relentlessly hospitable people even when they vehemently disagree with you.

If J.R.R. Tolkien's Middle Earth were a real place, its heroes — the humble, agrarian Hobbits of the Shire — would live in Iowa or Ireland. And being partial to good soil, I imagine Hobbits would prefer the rich, black loam of Iowa over Ireland's rocks and rushes.

It was, of course, Hobbits who saved Middle Earth from death, darkness, and unimaginable evil. Perhaps in the New Climate Era the down-to-earth farm people of places like Iowa and Ireland hold the key to our species' survival on this beautiful planet we call home.

So, I write this book to chronicle a journey of seven million steps, sounding the alarm that we have unleashed great evil upon our collective home, attempting to rouse a nation out of apathy, challenging everyday people to extraordinary action.

Perhaps through sharing this story of walking a great distance against so many external odds, while wrestling with internal struggles that bubbled closer to the surface as the miles ticked away under foot, I can make one more contribution to the future we all hope to enjoy.

CHAPTER 1

FAILURE

"The loneliest moment in someone's life is when they are watching their whole world fall apart, and all they can do is stare blankly."

— F. SCOTT FITZGERALD, ***THE GREAT GATSBY***

    I wake to a light rain misting my face. The caress of a moist leaf brushes my cheek as it falls from a tree. Another leaf drifts toward me, lilting back and forth, the water droplets on its surface glistening in the soft glow of a nearby streetlamp. Smiling, I fight to stay asleep, and dream I'm watching a half-naked fairy princess ride another leaf to the ground like a silent green glider.

    But consciousness fights me and wins. I feel the hard, cold ground against my spine taunting the battered muscles of my arms and legs. The soothing white noise that was a bubbling stream in my dream fades into the shrill blight of heavy traffic.

    The pulse of oil-powered steel a mere twenty feet away reminds me of everything that is wrong with this moment, reminds me of my utter exhaustion and aloneness. The traffic seems to shriek, "You've failed. You've failed. You've failed."

    Have I failed? It certainly feels like it. To have come more than 3,000 miles on foot only to collapse under a stand of trees along a highway somewhere west of Washington, DC, does not ring of success.

    For eight months, the Great March for Climate Action had set its sight on the White House. Every day our destination drew 15 miles closer. I'm now less than 20 miles away, and the realization sinks in that I'm not going to make it.

I've made my share of mistakes during this transcontinental odyssey. Some of them flash before my eyes as I lie under the trees, the mist now bringing a chill to my body. But it seems I've saved the worst mistake for today.

After an exceptionally poor night's sleep, I miscalculated the availability of food and water along the C&O Canal Trail. I ran out of water at the ten-mile mark and walked 22 miles with nothing to eat. On top of that, the plantar fasciitis that first troubled me in Nebraska returned with a vengeance.

Compounding my problems, halfway through the day my phone died — the phone that was instrumental for conducting March business and, even more important, for navigation. Unable to find a power source along the trail and unable to access a map, I had no idea where I was.

Now I lie under these trees trying to remain present, assessing my condition and options. I want to remain invisible. I doubt drivers on the nearby highway can see me. Yet I wonder about the people in the mansion through the trees just beyond the fence. Perhaps the mansion's security cameras have been monitoring my presence all along. Perhaps the police are already on their way to arrest me and charge me with trespass.

I lie still, like a wounded animal hoping to avoid detection from a real or imagined predator. I try to stay warm yet feel unable to move. The reality of my isolation presses ever harder against my chest, against my heart.

I had left the other marchers a week ago after conflicts that had grown so virulent and tiresome that I desperately needed a break. I needed to walk in peace before rejoining the group for our final hike to the White House. But now I miss the camp, despite its dysfunction. I miss the familiar comfort of my tent. I half-hope, half-imagine, that Sarah Spain, the March's gregarious logistics director, will drive by, sense my presence, and pull over to rescue me.

Sarah has certainly worked her share of miracles on the March, instantly connecting with whoever stood between us and whatever we needed. In an Arizona bar, Sarah's cowboy hat and eagerness to pose for a photo with a shotgun scored us a rodeo-ground campsite. In New Mexico, her charm with a commune of aging hippies landed us foot massages and a delicious meal.

Whether it was campsites, food, vehicle repair, or showers, Sarah always found a way to deliver.

But today, there would be no Sarah-magic.

I feel the sting of loneliness on a deeper level, too. Two years ago, I met Grace, the woman who I was certain was my soul mate. Grace and I talked often but weren't able to spend much time together. When I left Des Moines in February, I was confident that would change, certain that when I returned from the March, Grace and I would begin our life together. But what transpired during our months apart sowed doubt.

A desperate sadness sinks in and I imagine its weight will push me deeper and deeper into the ground.

A truck roars by on the highway, wailing on its horn, rousing me from my lethargy, transporting me back to the reality of my place under these trees dripping with mist, coldness, and irony. I ask myself, "How the hell did you come so far to end up here — battered, alone, defeated?"

CHAPTER 2
WHY NOT A MARCH!

"Very few people on earth ever get to say: 'I am doing, right now, the most important thing I could possibly be doing.' If you'll join this fight that's what you'll get to say."

— BILL MCKIBBEN

In February 2007, Bill McKibben came to Iowa State University where he spoke to a packed house about climate change and the loss of community. His message was a blend of warning and hope. "Americans will be happier if they return to the 1950s lifestyle of eating together as a family, talking with neighbors and carpooling to work," McKibben exhorted.[1]

Consistent with his call to simplicity, McKibben joined a group of us afterward for chili and corn bread. We sat on stark benches under dim lights at a long table in one of the campus dining rooms. I guess you could say it was very 1950s, but my mind got stuck on the dining scene from *Oliver*.

I have no stomach for self-absorbed celebrities, and I've met plenty. Joe Biden came to mind as I sat across from McKibben. Biden had called to seek my endorsement in his campaign for President in 2006, the year I ran for governor in Iowa. As we settled in for a beer and a game of pool, Joe asked about my campaign. Twenty seconds into my response, Joe jumped in, and for the next hour the conversation

---

[1] Speaker cites loss of community, by Karla Walsh, Iowa State Daily, February 20, 2007

was all about him. Nice guy. Captivating stories. Lousy listener. Biden won the game of pool but lost my vote.

I can't speak to McKibben's pool game, but when it came to conversation, he was the opposite of Biden. He was down-to-earth and unpretentious. As we blew on spoonfuls of hot chili and corralled wayward crumbs of cornbread, Bill mostly listened and asked questions. He was genuinely interested in what students were up to and curious about my campaign for governor.

I told him it had been edgy, quirky, cobbled together. We didn't pay a penny for our campaign headquarters — an unused house in Des Moines' inner city.

Well, not completely unused. The house was infested with possums who were outed one day by a campaign worker's shriek as a furry head appeared in the heating vent. Over the course of the next week, three possum-squatters were captured, evicted, and relocated to more upscale digs on the rustic fringe of suburbia.[2]

On Election Day (June 6, 2006 — or 6-6-6, for what that's worth), I finished a respectable third place, crediting the campaign's strong performance to its grassroots nature, solid support among suburban possums, and firm positions on controversial issues. I'd talked about money in politics, economic justice, marriage equality, and climate change.

"Climate change isn't an issue," Bill interrupted, without any haughtiness or condescension.

"What?" I responded. "You're the climate-change guy. How can you say, 'climate change isn't an issue'?"

"It's not an issue," Bill said. "It's a crisis."

That caught me off guard. I thought about it a lot during the ride home. Over the next few days, as I read more about climate change, the truth of McKibben's words weighed heavier and heavier upon me. Humanity was indeed facing a crisis like none other in history. A world of hurt was coming our way, and if we continued a lifestyle powered by fossil fuels, Earth could be rendered uninhabitable for our species and most others.

I began to prioritize climate action in my work, but struggled for a

[2] http://qctimes.com/news/opinion/editorial/columnists/maverick-fallon-knows-anything-can-happen/article_2e7fa12b-50cf-52fb-b9c4-3d78e6588e58.htm

long time with what big contribution I was called to make. Unexpectedly, one blustery day in February 2013, it came to me: "Why not a march!"

In that moment, the Great March for Climate Action was born. A thousand people would walk 3,000 miles across America. Our commitment and sacrifice would draw others into conversations about climate change. We would be the Paul Reveres of the movement, sounding the alarm and inspiring tens of thousands of people to take action.

I started raising money, writing press releases, putting together the outline of a plan. The vision needed structure and discipline. I hired Shari Hrdina, an exceptionally gifted tactician who enjoys all the aspects of organizing I despise. Forms, spreadsheets, and paperwork make me apoplectic. They make Shari giddy.

Our destination was clear: Washington, DC, where ultimately the most far-reaching decisions about climate change would be made. But the starting point was negotiable: somewhere on the West Coast between Canada and Mexico. I spent weeks pouring over maps, studying terrain and elevation, analyzing weather charts, and talking with local authorities up and down the West Coast. The most enthusiastic supporters stepped forward in San Francisco and convinced us that the Golden Gate Bridge should be our iconic starting point.

As we plotted our path upward through the Sierra Nevada in late winter and early spring, the words "Donner Party" kept coming up in conversation. The Donner Party, of course, was the unfortunate band of settlers who, trapped in the Sierra Nevada in the winter of 1846-47, survived by dining on the Party's less-fortunate members.

Further analysis of weather trends, conversations with locals, and a proliferation of uncomfortable jokes about marchers eating marchers compelled us to abandon what had been no small piece of work and seek a more southerly route.

After another round of deliberations, I settled on Los Angeles, which made so much sense for so many reasons. Los Angeles had been the starting point of the Great Peace March. Southern California was normally comfortable and dry in early March, and beginning there would keep us in the desert and south of the Rockies until warmer weather.

With our start and end points determined, everything in between was up for grabs — everything except Nebraska.

Of all the places in the country threatened by the aggressive build-out of fossil-fuel infrastructure, central Nebraska was ground zero in the battle to stop TransCanada's Keystone XL pipeline. In ten of eleven states we would walk through, the main event would be a rally or march in a large city. In Nebraska, the main event would be a rally in a cornfield at the point where the March crossed the proposed path of the Keystone pipeline in the heart of the Sandhills.

The intensity of organizing mushroomed over the summer. We needed more staff fire-power to pull the remaining pieces together. I hired my friend Jimmy Betts as program director. Jimmy had signed up to march and volunteered in our office from time to time. He'd participated in some of our training walks, sometimes playing his violin during the two-mile stroll around Gray's Lake in Des Moines.

Jimmy's work history was, well, interesting. He taught Qigong and meditation and had held a "real job" for a short time. Occasionally, he'd give blood to earn cash or, worse yet in my opinion, sign up as a human guinea pig for some experiment to test a new food or drug. I told him I'd rather eat weeds and live under a bridge than earn money like that.

These days, Jimmy mostly traveled between Texas and Iowa, hanging out with friends, sharing his passion for ancient wisdom and disciplines, sometimes for pay, sometimes not. I had a hard time imagining that he'd stick with us for the duration. But I also felt there was some great, untapped potential in Jimmy that could be unleashed through his full-time involvement with the March.

One of my guiding principles is that, if you ask people to do a difficult thing, you have to be willing to do it yourself. Whether it's a hunger strike, civil disobedience, or a march, the history of progressive movements is littered with the fleshy bodies of those who challenged others to greatness while avoiding the hardest work themselves. With this in mind, I initiated an aggressive fitness program.

When I began training in July, even a three-mile walk left me sore for days. By the end of November, I was able to log 25 miles a week. I grew increasingly confident about my ability to walk long distances and wondered if I might be able to manage all or most of the March — though a haunting history of back trouble left plenty of doubt.

Yes, a series of back injuries in my 20s and 30s had debilitated me to the point where walking more than a short distance was impossible. My right leg had atrophied badly and I was told my back would never again be fully functional. It was indeed a miserable 15 years. The nerve damage on my right side was so bad I could sit for only 10 to 15 minutes at a time. During college classes, I stood in the back of the classroom with my right leg up on a chair for relief. I had to lie down during car rides and couldn't lift or carry my kids.

I was a wreck, but had come to accept that this as my lot in life. Doctors provided no relief, each misdiagnosing the problem according to his specialty. I was told I should apply for disability, but couldn't bring myself to admit I was disabled and didn't want to depend on public assistance. After five misdiagnoses, a friend studying to be a physical therapist correctly diagnosed the problem as a slipped disk.

In my mid-30s, with the help of other physical therapists, I began to heal. I started biking. My right leg recovered its strength and the atrophy gradually disappeared. I discovered the importance of flexibility and core work and, over time, the muscles in my abdomen and back grew stronger. It took years, and I knew I would always have to be cautious, but eventually I regained the ability to live a more physically active life.

Two months before the scheduled start of the March, I was walking 35 miles a week. I felt so good about my progress that I thought I had a shot at walking the entire distance, coast-to-coast.

Then in early February my back inexplicably spasmed. I was confined to bed for three days. After a week of gradual recovery, I began walking a few miles at a time. Since my late-30s I had come so far toward restoring and maintaining a healthy back, toward living a normal life. Why did this happen? Why now? I was beyond disappointed and certain I had ruined any chance I had of marching much.

Still, I felt enthusiastic as I prepared to leave for Los Angeles. Despite many obstacles, the March had come together reasonably well.

As my departure from Des Moines drew closer, I felt a heightening of that primal tension — the lure of an adventure into the unknown versus the familiarity and comfort of home. Overriding both was a strong sense of duty to sacrifice for the life-and-death cause of climate change.

My piano would sit idle for the next eight months. This would be the first year in thirty I wouldn't tend a garden. I would miss my home, my community, my friends. Especially, I would miss Grace.

What I craved above all else was a quiet, settled life with Grace — the woman I loved deeply — to one day be with her in a home alive with love, food, and music. That dream was shoved aside by the call to service. The path into the unknown had been charted. Its first steps led out my back door into a cold February night, away from the home I wouldn't see again for a long time.

CHAPTER 3
SECOND-CLASS TRAVEL

"Thanks to the Interstate Highway System, it is now possible to travel from coast to coast without seeing anything."

— CHARLES KURALT

The three worst inventions in the industrial history of man (yes, blame men) are the television, the nuclear bomb, and the automobile. I've been mostly successful at avoiding television, to the point of being culturally illiterate. Someone once asked what I thought of *Downton Abby* and my response was, "Huh. I didn't know Des Moines had a monastery."

Regarding nuclear bombs, perhaps simply through dumb luck, I have so far managed to avoid them. Given that Earth is home to 15,000 nuclear weapons, 1,800 of them on high-alert and capable of being launched in a matter of minutes, dumb luck is a quality I share with all of humanity. Let's hope it holds. Better yet, let's hope the nuclear nations of the world come to their senses and disarm.

Regarding the automobile, my track record is mixed.

During America's 240 years of nationhood, we've built remarkable, even beautiful infrastructure. We've also built spirit-crushing rubbish that would have made Josef Stalin proud. The Interstate Highway System is of the latter genre. The ascendency of the automobile as the sole form of transportation for most Americans was enabled and solidified by this system.

It's as if the Interstate's designers said, "Let's create the most soul-sucking, isolating travel experience imaginable, then underfund or eliminate every other form of transportation so people have no choice but to use it."

In this, the designers were eminently successful. Not counting time spent as passengers, Americans drive an average of 280 hours each year, about 24 percent of that on Interstate highways — or "freeways," as we've been conditioned to call them.[1]

"Freeway" is one of the most brilliant linguistic coups of modern times, more impressive even than changing the name of the War Department to the Department of Defense, or calling civilians killed in war "collateral damage." Calling something "free" that cost Americans $114 billion to build and hundreds of billions more to expand and maintain is social engineering at its finest.

Because it's "free" and, for many people, the only viable transportation option, few Americans question the necessity for ever-bigger, wider swaths of asphalt and concrete.

Unlike televisions and nuclear bombs, I do own a car. She's a rusty, battered old Subaru named Beast that my friend and former legislative colleague, Bill Witt, sold to me for a buck. My previous car was another rusty, battered old Subaru, also named Beast, also sold to me by Bill for a buck.

(And no, Bill isn't a Subaru dealer. He's a photographer and generous to a fault. So ask Bill for a photo of a Subaru, but don't pester him about whether he's got a real one kicking around that he'll sell you for a buck.)

My first Beast and I shared seven years of car-man bliss before she dramatically exited this life in a glorious burst of smoke at 60 miles an hour in heavy traffic. True to the end, Beast guided me to the road's shoulder unharmed as she bellowed her final dying breath. In my present life in Des Moines, thanks to an old bike and good walking shoes, I can go a week or more without commissioning my new Beast into service.

Yet as a state lawmaker and candidate for governor and Congress, I was imprisoned in a car for as many as 30,000 miles a year. My occasional furlough was to escape the Interstate and meander backroads homeward, annoying my children and frustrating my staff. As far as I could gather, these furloughs were bothersome to my passengers

---

[1] http://newsroom.aaa.com/2015/04/new-study-reveals-much-motorists-drive

because travel in America is not about the journey. It's about getting someplace as quickly and cheaply as possible — someplace where life begins once you're done squandering time on travel.

A backroad is a quieter, simpler place, designed with priorities other than speed and size, built with aesthetic sensitivity to the vagaries of both land and human habitation. On backroads, you're closer to the land and able to appreciate features that don't even register from a car barreling down the Interstate. On backroads, you often wind your way through small towns and experience fleeting impressions of those towns' unique characteristics. You're more inclined to stop, which is good for your spirit, good for your posture, and good for the town's often struggling economy.

A backroad can be beautiful to drive — and as I was soon to discover, even more beautiful to walk.

It's not possible to live without occasionally compromising one's principles. I have driven and will drive again. I have flown and will fly again. But it was impossible to justify driving or flying to the start of a cross-country march whose mission was to move America beyond fossil fuels.

On February 23 of 2014, I board Amtrak for the 48-hour train ride from Iowa to Los Angeles for the start of the Climate March. My gear consists of two duffle bags, one containing my camping gear, the other clothing and personal supplies. I carry my walking stick and a small satchel given to me in 1995 by Sumitra Kulkarni, the granddaughter of Mahatma Gandhi. With the addition of a liner and four pockets, the satchel is the perfect size for carrying what I need for the day's march: water bottle, wallet, headset, lunch box, sunscreen, and phone charger.

As I settle into my seat on the train, I think of how, beyond its stated purpose of sounding the alarm about the climate crisis, the March will serve as a counter-cultural statement about the value of walking as transportation.

Whereas walking doesn't even register on Americans' list of transit options, train travel in America is second-class transit at best — at first-class prices. I drop $700 on my one-way ticket to Los

Angeles, twice as much as I would have paid to fly. But factoring in that this price covers six meals, my "roomette" in the sleeper car, and the lowest possible carbon footprint, it's a worthwhile investment.

Surprisingly, the food on the train doesn't suck — nor do the social aspects of dining. I appreciate how the wait staff never ask passengers entering the dining car where they want to sit. They simply show you to your table, always with another passenger, probably one you've never met. This has the effect of encouraging something foreign to modern American eating habits — a dining experience that melds mostly satisfying food with, if you're lucky, mostly satisfying conversation.

My roomette in the sleeper car is cramped but manageable. It's barely larger than the tent I'm about to call home for the next eight months. I'm impressed that such a tiny space can meet so many human needs.

As the train rolls westward toward the Rockies, we fall further and further behind schedule. Freight haulers own the tracks, and we sit and wait while a cargo train passes, often hauling coal and oil. Once — yes, just once — we pass a train stacked with wind turbine blades. "At least that's a start," I muse.

Free of the need to grasp a steering wheel, train travelers have ample time to read, converse, roam about or, in my case, work. There is still so much to do before the start of the March, which is now less than a week away. Like my campaigns for governor and Congress, I'm constantly raising money. A slew of logistical details still have to be addressed, and fewer than half our California campsites have been secured.

Internet and phone service are intermittent, affording me occasional time to think and reflect. I daydream as I study the changing landscapes that roll by, wondering if I'll soon walk across this particular stretch of country. I feel I'm being offered a preview of some of the beauty and challenges that lie ahead.

I have time to think about my life, too. At night, before bed or when I can't sleep, I stare out the window of the car at the silhouettes of buildings and landscapes in the distance, or sometimes very close, too close to see with clarity, merely a blur and a flash of shadow and light. When I look at the window itself I see my reflection, see a man who looks confident and ready to tackle a great challenge, but whose eyes belie more than a trace of sadness and loneliness.

There are times when the clacking of the train against the tracks simulates the sound of an old movie wheel rewinding. I stare at the man in the window. Why this sadness? Why this loneliness?

I think about my two failed marriages. Plenty of sadness there. More than anything, I think about Grace, how we fell madly in love, a love like none I've ever known. I miss her. I want to be with her so badly, and find comfort knowing that when the March is over we'll begin our lives together.

CHAPTER 4
POST-VICTORIAN MAN

"If you build your life on dreams, it's prudent to recall, a man with moonlight in his hands has nothing there at all."

— *"TO EACH HIS DULCINEA,"* **MAN OF LA MANCHA**
(lyrics by Joe Darion)

Despite occasional spells of doubt, I nurture the notion that as a man slogs through the tasks and tumult of his daily life, he eventually meets the woman of his dreams and they both know with unshakable confidence that they're meant to spend the rest of their lives together. Admittedly, I've met only a few such couples in the flesh, though they can be found in abundance in fictional literature, the bulk of it pre-dating the 1950s.

For better or worse, somewhere in the middle of the last century, there occurred a seismic shift in how American males regarded romantic relationships. The love-intoxicated, idealistic man-hero of Victorian times sobered up. He accepted the fact that divine forces weren't simply going to guide him to his true love, where recognition of their predestined union would be mutual and instantaneous.

Post-Victorian Man — P-V Man, we'll call him — learned that love is more than chemistry and predetermination. With the strength of character and clarity of purpose that have come to define our gender over the millennia, P-V Man knew that if he were to reel in an adequate life partner it would, alas, require time and effort. He wisely pivoted his strategy, seized the initiative, and began to spend prodigious amounts of time in bars.

In that primal but pricey tabernacle of relief and self-delusion, after often prolonged efforts to silence the growing realization of his loneliness with excessive quantities of beer, P-V Man would indeed meet the woman he would marry — and later divorce. Sadder still, he would discover that beer-therapy for a love-sick heart was snake oil. This moment of enlightenment would, of course, lead to the consumption of even greater quantities of beer.

Whether one takes the plunge as Victorian Man or P-V Man, the raging waters of human love are complex. They can be breathtakingly beautiful as well, but so full of unseen hazards they are best navigated with the senses unimpaired.

For confirmation of this truth, regard the mighty hero Odysseus, who had the presence of mind to have his crew bind him to the mast of his ship. Thus bound, he could hear the seductive song of the Sirens without losing his mind and throwing himself into the sea like so many Greek sailors before him.

By virtue of either my Catholic upbringing, idealistic nature, or perhaps a genetic defect, I totally missed the P-V chapter of American-male history. Through five decades and two divorces, my unfettered confidence in the reality of the Victorian love model remains rock solid.

Mostly.

Dusk melts into the dark of a central Nebraska night. The ambling movement of the train and clacking white noise of her progress along the tracks induces a reflective calm.

I think about the history of my half-century of life on Earth, a chronicle of failure punctuated by occasional success. Nowhere had I failed more gloriously than in the realm of love. I take note that it was here, crossing Nebraska 32 years ago in a Greyhound bus, that I met my first wife.

That experience was a true Victorian-Man moment. I was visiting a Shoshone spiritual encampment in the Nevada desert on my way to an Ojibwe reservation in northern Wisconsin to direct a choir and teach music. I missed the connecting bus in Cheyenne, Wyoming.

When I boarded the next available carrier I was pleased to snag two seats to myself. Across the aisle, an attractive young woman had also laid claim to two seats.

The last passenger to board the bus was a young man, defiant and drunk. He sat next to the woman and pulled out a bottle of whiskey. The driver threatened to throw him off the bus. The man grudgingly parted with his beverage, turned to the woman, and commenced to rage against the oppression inflicted by bus drivers on innocent, well-intentioned drunks.

As always in those days, my back gave me trouble. When traveling by bus, I would walk up and down the aisle for relief. On this trip, I sauntered to the back to discover a game of poker in progress.

I am neither a poker player nor a gambler, but the card game provided an excuse to avoid sitting. I accepted an invitation to join. With only $100 to my name I thought, "What the hell, if I can slowly blow 20 bucks and avoid the discomfort of sitting, it'll be worth it."

Four hours later, I was $20 richer and had gained both the respect and contempt of three gamblers who were, without a doubt, far better than me at poker. When I returned to my seat, the woman who'd been stuck next to the drunk had moved over to my seemingly abandoned space.

Her name was Kristin, and as our bus rumbled across the prairie on a hot summer's night, we got to talking. We were both musicians. She played the harp. I had never seen a harp, let alone met a live harpist. I was enchanted by both.

It was only the year before that I had heard the harp for the first time while staying at a convent in Cairo, Egypt, during my travels. The head nun had given me a cassette recording of harp music. I was instantly drawn in by the instrument's sweet, delicate sounds — to me, the tonal equivalent of lily-of-the-valley. I would listen to the tape over and over while heaving pillows into the upper reaches of my room's 12-foot ceilings to try to kill the exceptionally large, fast-moving mosquitoes that hovered above me, laughing and placing bets, I imagined, on which of them would get to inflict me with malaria.

"The harp!" I said with excitement. "That's my favorite instrument!"

Kristin looked at me suspiciously and asked if I'd ever heard one. I could've told her about the cassette I'd listened to in Egypt, but instead muttered, "Well, no, not really."

Awkward pause. "So, what do you play?" she asked. I could have said the piano music of Chopin or classical guitar or Bach on the organ. Instead, I blurted out, "The accordion!" She stared at me blankly.

It got worse. Kristin showed me the hand motion one uses to play a harp. I responded enthusiastically, "Wow! You'd be good at milking cows."

That seemed to me a high compliment, similar to the high compliment I had paid two girls named Wendy in kindergarten, to whom I proudly announced one morning that I'd named my dog after them.

The cow comment went over with Kristin about as well as the dog-naming honor had gone over with the two Wendys. Instead of, "Why thank you, yes, I would love to milk a cow someday and am so glad I have unwittingly developed this skill set," the offended, classically-trained harpist whom I had only just met slapped me.

I can't say for sure, but that might have been the moment we fell in love.

As a married couple, Kristin and I lived in the inner city (yes, Des Moines has one) in a fixer-upper we bought for $1,500. Our neighborhood was the poorest and most racially diverse in Iowa. Economically, we were well-matched with our neighbors: during our first eight years together, Kristin and I had a combined annual income of $12,000 to $20,000.

Yet those were the happiest years of my life with two children, dozens of housemates who became our extended family, lots of visitors, pets, chickens, a big garden, and great neighbors.

Kristin and I were very much in love. Yet my mind and heart remained unsettled. I had not been ready to marry. With increasing frequency, I reasoned that I needed to be with a partner who stood side-by-side with me in the fight for justice. Especially when my work involved travel, I felt a profound loneliness that reinforced this rationalization.

I convinced myself that the ideal partner was out there somewhere. While pleasant on the surface, my marriage with Kristin was afflicted with a strong undercurrent of dysfunction that I wasn't able to address and that we almost never discussed.

Lacking the tools to deal with this, and being foolishly unwilling to seek professional help, I did what toddlers do. I acted out — and in a way so inconsistent with my life-long commitment to honesty that it shocked a lot of people and still shocks me.

I had an affair.

Staring at my reflection in the window of the train passing through the farm fields and small towns where I first met Kristin, I'm reminded that the shame of that failure will stick with me the rest of my life.

My second wife, Lynn, was the woman with whom I had the affair. We met in 2005 when she worked on my campaign for Governor. Inconsistent with the Victorian model, I felt no romantic attraction to Lynn. But we worked well together, and I convinced myself that this was more important than a strong attraction. Lynn satisfied my yearning for a female companion in the day-to-day grind of public work. Over time, I grew to love her and was convinced that this was the partnership I needed to fulfill the work we both felt called to do.

My marriage with Lynn was embarrassingly short-lived, lasting less than two years. When the appeal of working with me wore off, she moved on — and moved on quickly.

I had done everything I could to shift the focus of our work to revolve around around her. But Lynn found more and more reasons not to work with me or spend time together. She said she didn't want to do "menial" work, but took a job doing menial work with a man with whom she would frequently travel, often overnight. I suspected it was more than that. It felt like karmic payback, a taste of what I had inflicted on Kristin.

The words Lynn chose when she told me she wanted out will forever stay branded in my mind: "I don't think this relationship is sustainable." I was floored, and could not have conceived a more nuanced way to say, "I want a divorce."

When I married the first time, I did so doubting it could last, even though Kristin and I were genuinely in love. When I married the second time, I was certain it was for life, even though our connection was based not on strong attraction, but on an appealing work relationship.

As I thought about my two failed marriages it struck me that, perhaps at least for me, the Victorian Era had it right regarding human love. Perhaps with Kristin I had been with the right person and I'd simply let my rational mind muck it up.

Adjusting to life as a single man was painful. I was in a rush to meet someone with whom I could establish the long-term, committed relationship I craved. But dating proved to be a disaster. One relationship ended when my lover literally salted the earth in my garden. Another ended when the woman hurled a large, rotten squash at me and scored a direct hit. Other attempts at dating fizzled for one reason or another, not always involving vegetables.

As my train takes me closer to the start of the March, more than anything, I'm hungry for a partner and eager to begin my life with Grace later this year. But I also crave community — people with whom to share meals and a physical living space. This is one of the March's ancillary appeals: instant community — a passionate group of climate activists thrust into a mobile village doing something difficult and important for eight months.

As the train nears the end of the line, I'm excited to meet my new family and begin our all-important mission.

CHAPTER 5

BASE CAMP

"Our lack of community is intensely painful. A TV talk show is not community. A couple of hours in a church pew each Sabbath is not community .... Without genuine spiritual community, life becomes a struggle so lonely and grim that even Hillary Clinton has admitted 'it takes a village.'"

— DAVID JAMES DUNCAN

As a boy growing up on the New England coast, it would take me 20 to 30 minutes to ease into the limb-numbing waters of the North Atlantic. Some kids would plunge into the icy sea in a matter of seconds with utter disregard to both system shock and sand sharks. These kids scored points for efficiency and time management, but I had the corner on caution and prudence.

It thus made sense to me to ease marchers into our lives as tent-dwelling nomads with three days of civilized camping, modern amenities, and mowed grass. Our California coordinator, Ki Coulson, found the perfect place at Camp Fire Long Beach, less than twenty miles from our starting point at the Port of Los Angeles in Wilmington.

By the time I arrive at our base camp early on the morning of February 26, a handful of marchers have already pitched their tents at strategic high points (in case of rain) and under large trees (in case of sun) along the long, narrow swath of lawn.

People gradually trickle in as our camp grows over the next three days. Most of us know each other only through our online profiles.

Yet there is immediate camaraderie and palpable enthusiasm — summer camp and that long-awaited family reunion fused with mission and purpose.

In addition to briefings by staff and guest experts, final preparations for the March include some major challenges and obstacles. Our environmentally friendly "Eco Commodes" haven't arrived, and won't for another week. Fortunately, during most of our six-day, 80-mile trek through metropolitan Los Angeles, we'll avail ourselves of restrooms at some of the thousands of fast food restaurants along the way — possibly the only time in my life I'll be grateful for McDonald's.

Our water tanks haven't arrived either, being mysteriously diverted to the wrong location. It takes a week, a ridiculous number of phone calls, and many miles of driving to retrieve them.

The coolers we purchased for perishable foods are highly efficient, but it becomes clear during the first two weeks that we'll need a commercial refrigerator. Ice for the coolers is expensive and will be difficult if not impossible to find when we cross deserts and mountains.

We're still short two camp sites for the first week and, worse still, the Bureau of Land Management (BLM) won't allow us to camp anywhere in the Mojave Desert during our nine-day crossing. Ki and I puzzle extensively over how to deal with that problem. But more important matters demand our immediate attention.

Matters such as eating.

The rules for food trucks in Southern California are jaw-droppingly stupid. I'd secured a $20,000 donation to purchase a food truck only to learn from an LA County inspector that we're required to quarantine the truck every night in a designated commissary. We explain that this is impossible because our mobile camp will be 15 miles farther east every day. That doesn't matter, we're told — the law's the law.

We tell the inspector we're not going to sell food and that marchers will do all the cooking and food prep. Mistake! Now we'll have to pay $150 per person for each marcher to complete a training program to become a "certified food handler." We also learn that the March needs a $600 vendor's license from *each* of the twenty-or-so cities we'll walk through during our one-week crossing of LA County.

Compliments of my extensive tenure as a state lawmaker, I was familiar with government bureaucracies run amok. But this beat all.

After much frustration and frequent banging of heads against walls, we decide to do what American patriots have done since before the inception of this great nation — ignore the law and innovate our way around the legal obstacles thrown at us by the enemies of freedom and common sense.

The March had purchased a 35-foot straight truck to haul gear. Why not cordon off a section of the truck for our kitchen? Just tuck it in there somewhere among the gear and hope government inspectors are too busy harassing others or enjoying long coffee breaks to bother us.

Sure, hauling food and water in the same vehicle as tents and personal gear was less than ideal. But the plan would get us through California. In Arizona we'd rent a smaller truck to serve as our mobile kitchen for the duration of the March. That option worked almost as well as a food truck — at one-tenth the cost! I have no doubt that county health departments would have found numerous violations. But we never ran into an inspector and, as far as I'm aware, no one ever got sick from eating food prepared by our amateur chefs and frequently barefoot, un-showered marchers.

Another problem stood in the way of marchers and meals. The person in charge of assembling the necessary kitchen supplies had dropped the ball. Three marchers rose to the challenge of outfitting our bootleg kitchen. Marie Davis, Debaura James, and Lala Palazzolo took by storm the local thrift shops and a nearby grocery store. In less than a day, on a tight budget, they purchased the food and equipment needed to prepare and serve meals.

On our second night in Camp Fire, with our clandestine food-prep operation cranking out dinner in the back of the gear truck, Marie made a simple but delightful pesto dish. Breakfast and lunch the next day were basic but adequate. That evening, Jeffrey Czerwiec made an Italian lentil-kale soup that was well received.

Thanks to the diligence of a handful of marchers and our collective ability to navigate around pointless rules and regulations, we had, at least temporarily, solved our food-prep problems in less than 48 hours.

The challenges of our final preparations were significant, but they only made us stronger, more determined and more united as a community of marchers. Even the steady rain during most of our time at Camp Fire didn't dampen our enthusiasm.

Tomorrow, the March will begin! Tonight, through a light drizzle, I admire the tents assembled on the lawn beyond our dining area. I think of what a contrast our community — which we call "One Earth Village" — presents to typical life in America. Our social structure is so counter-cultural as to be almost inconceivable to most people. Sure, privacy and personal space are important. But feeling connected to others is an essential element of the human condition, one that has largely been destroyed in modern-day America.

Just as our car-centered transportation system isolates us when we're mobile, the design of our homes and neighborhoods isolates us when we're stationary. When walls that separate us from each other are fashioned in wood and stone, we're physically and psychically cut off from the world. When one's walls are the thin fabric of a tent, when most of life's activities are conducted in shared space, an entirely different social energy emerges.

In America today, it's quite possible, perhaps even the norm, not to know most or any of one's neighbors. The large porch that served in prior times to link one's private space to the common space of the street has been replaced by the large garage. Instead of sitting out front we now sit on back decks surrounded by tall privacy fences.

Isolation-by-design is even worse inside our homes. The living room's "entertainment center" has replaced the public theater for plays, movies, and concerts. Meals, even groceries, can be purchased by phone and delivered to one's door. Nearly all shopping can be conducted online. With a small investment in a treadmill and weights (delivered to one's door), a solid program of physical exercise can be managed without ever stepping outside.

To call such a set-up a home is disingenuous at best. More accurately, it is a pricey, private prison cell. Clusters of these cells are called neighborhoods, and the interactions between one cell's inmate and the neighboring cell's inmate are typically minimal or nonexistent. In fact, inmates at real prisons have more contact with each other than those imprisoned in most modern neighborhoods.

To say that this arrangement is problematic is an understatement. When future generations of sociologists and psychologists analyze why there was so much violence in America in our time, I'm certain they'll conclude one of the primary reasons is that human beings cannot be healthy and normal when disconnected from each other and the natural world.

From a political perspective, when we're isolated we're more easily duped by propaganda. When we're confined in subdivisions, when our main source of information and ideas comes through a screen, not through direct dialogue with people facing similar life challenges, then the collective "we" is weak, easily exploited, eager to scapegoat those who are "different," inclined to buy the lies of, say, a Donald Trump.

Despite all the discomforts of living in a tent for eight months, one of the values of One Earth Village is that we won't be lonely or disconnected. As I look out over our encampment, I feel as if I'm about to go on a blind date with 50 people. I know we'll have conflicts, but I also know that many, if not most, of these people will remain my friends for the rest of my life.

The rain lets up. Someone has lit a campfire. Music and conversation are underway. I join my new family gathered around the fire — the primal ancestor of the television — and take a turn picking a tune on the guitar. As I play I feel someone behind me. Hands begin to massage my shoulders and neck. It's Lala, and her hands feel good. This whole experience feels good. This connection with people in a simpler, more basic living arrangement, united by a common purpose, drives us to attempt what most people would never dream of doing.

The massage winds down as I come to the final notes of my song. "There's more where that came from," whispers Lala, loud enough for others to hear. Across the fire, I catch a glimpse of two of my Iowa friends, Karin Sandahl and David Thoreson, raising their eyebrows a bit, a fleeting, discreet Midwestern smile ever so slightly passing their lips, as if to say, "What was *that* about?"

I smile. Our time at Camp Fire has been intensely busy, but euphoric. In one year, staff and volunteers have pulled together the essential structure to propel 50 people on a historic 3,000-mile march across America. We're a smaller band of marchers than I'd hoped for. But we're eager to begin an effort that we hope will impact the national conversation on climate change in a significant way.

CHAPTER 6

IN THE SHADOW OF VALERO

> "The sooner we get started with alternative energy sources and recognize that fossil fuels make us less secure as a nation, and more dangerous as a planet, the better off we'll be."
>
> — SENATOR LINDSEY GRAHAM

On March 1, in what would become our normal routine for the next eight months, we crawl out of our sleeping bags before dawn, take down our tents, load our personal items into the gear truck, and pack away as many breakfast calories as our guts can transport.

Different from the soon-to-be daily ritual, instead of setting out on foot, cars and trucks haul us to Wilmington Waterfront Park where the March will begin in a few hours. Through bleary eyes as we drive along mostly quiet city streets I try to catch a glimpse of the sunrise. No luck. The clouds are thick, though at least the rain has stopped.

When we arrive at Waterfront Park, Native drummers have assembled for a sunrise ceremony to bless and honor our effort. A small, solemn group of locals gathers in a circle with marchers as the deep resonance of the drum beats the very pulse of Earth's heart and spirit. The high, piercing voices of the singers evoke Earth's sorrow at the ravages inflicted by her human children. I've heard Native drummers many times but this experience is richer, more meaningful than any I can recall. Some marchers are in tears.

John Abbe sobs convulsively. Miraculously, at one point during the ceremony, the sun pokes through the clouds for a few minutes as it rises in the East.

It's the only ray of sunlight we'll enjoy all day.

Ki and the SoCal Climate Coalition pull out all the stops to organize the rally, which draws 1,500 enthusiastic supporters. In addition to a big crowd and strong media presence, a documentary crew that had contacted me months ago is here. They plan to travel with the March all the way to Washington, DC.

Speakers and musicians take the podium as the towering infrastructure of the Valero oil refinery looms in the background. Valero has been accused of exacerbating the rampant blight and decay of a nearby neighborhood. Worse still, Valero threatens to refine greater quantities of oil, including tar sands oil, even as scientists announce that species are going extinct at a rate unprecedented since the days of the dinosaurs.[1]

The March begins. A group of students wearing dust masks and dragging a replica of a polar bear lead us into the street. I sense a collective rush of ecstasy among marchers as the first steps of this incredible 3,000-mile journey begin. I look at Sean Glenn, an amazing young woman who has decided to march the entire distance in silence. Sean is beaming. I beam back. Sean spent the weeks leading up to the March in Des Moines and, as we trained together, she'd become like a daughter to me.

We walk through largely Hispanic neighborhoods, the Pacific Ocean fading behind us and the skyline of Los Angeles opening before us. People line the street to cheer or wave from their homes. It feels more like a parade than a march. The sky had looked foreboding all morning, but so far there's no rain. "Maybe the clouds blew their wad the past three days," I say to myself. "Maybe we'll get through today without a drenching."

At the two-mile mark, our numbers drop to about 100 stalwarts who intend to go the day's distance — 19.4 miles through Los Angeles' industrial underbelly and some of its most impoverished neighborhoods.

About a mile later, the skies open to a flood of cold, driving rain that plagues us the rest of the day. The rain is like a strong summer

---

[1] http://www.biologicaldiversity.org/programs/biodiversity/elements_of_biodiversity/extinction_crisis/

storm in the Midwest, complete with thunder, lightning, and torrential downpours. The rain turns city streets into raging rivers that push against the curbs and eventually flood the sidewalks. We cross intersections wading through water up to our calves. The current flows so fast some marchers feel they could be swept away.

Our organized march quickly devolves into a struggle for survival. Marchers break off into smaller groups according to their pace or their need to stop for food, bathroom breaks, or to warm up. Some marchers are not even remotely dressed for the weather and show signs of hypothermia. Jimmy, Kim Foley, and I find one young marcher sitting outside a taco stand shivering uncontrollably. Kim is a long-time friend from Des Moines who works as a physical therapist. She checks the man's vital signs and drapes a sweatshirt around him while Jimmy calls one of our support vehicles to give the young man a ride to where he can dry off and warm up. I later learn that his story was fairly common today.

I could not have conceived a more difficult start to the Climate March. In drought-plagued Southern California, which had seen no rain for eighteen months, who could have imagined three days of rain before the start of the March and now this? By the end of the day, ten inches had fallen.

As the miles tick off, I keep expecting my back to scream, "Ok, dummy, that's enough." Yet, even though the day proves to be one of the most physically exhausting of my life, I'm able to go the distance. More than half the marchers make it, in fact. I'm with the final group — cold, soaked, but triumphant — who arrive at 9:30 at All Peoples Christian Church, where we'd been given permission to pitch our tents in the parking lot.

But the parking lot is now a small pond. Church officials agree to let us stay inside, a kindness we greatly appreciate. I barely have enough energy to wash up and crawl into my sleeping bag. We're 50 wet marchers crammed into an old gymnasium with acoustics that magnify every little sound. I'm a light sleeper and, despite my exhaustion, sleep doesn't come easily.

Eventually, I drift off and dream that I'm camping next to a pond, much like the parking-lot-turned-pond next to the church. But the pond in my dream is beautiful, surrounded by trees and grass, framed with a clear, blue sky. I smile to sleep in such an idyllic spot, enjoying the sound of frogs in and around the pond, all blending together in a pleasant symphony of the finest music an amphibious orchestra has to offer.

Then from the middle of the pond arises the croaking of a large bullfrog whose crass, disruptive voice drowns out all others. The sweet, symphonic performance becomes bitter, biting to my ears, jarring to my system. It forces me awake and I discover that the frog sounds are the snores of various marchers. The raucous, grating bullfrog is the snoring of Bob Cook, a 71-year-old marcher who suffers from sleep apnea. Tonight, we all suffer from Bob's sleep apnea.

The next morning I feel more battered than I ever remember. At 8.4 miles, today's march will be gentler, though I don't know how my body will respond after the beating it took yesterday.

On top of that, more rain is forecast.

CHAPTER 7

THE STREETS OF LOS ANGELES

"Los Angeles is 72 suburbs in search of a city."

— DOROTHY PARKER

As we prepare to set out, I see the church's old and stately exterior for the first time in daylight. The water that filled the parking lot last night has subsided, for the most part. A well-kept vegetable garden catches my eye. Although I know the growing season is much earlier here than in Iowa, I'm still amazed to see mature plants on March 2. I feel a twinge of sadness at this reminder that I won't plant a garden this spring.

The rain is not so bad today, merely an off-and-on drizzle. Having survived the previous day's deluge, we hardly notice. Today's comparatively short walk might have been easy under normal conditions. But blisters, sore legs, aching backs, and sleep deprivation make the day a challenge for nearly everyone.

We walk through LA's flower district to a to Sage Granada Park United Methodist Church in Alhambra, where the congregation rolls out the red carpet to an extent unexpected. Church members serve a unique blend of Japanese and Polynesian dishes followed by music from a 15-piece ukulele band.

Some of us set up tents. Others choose to sleep in the library, which after dinner becomes a massage parlor of sorts with marchers taking turns relieving each other's sore backs and limbs. Learning that Lala's offer of "there's more where that came from" is a quid-pro-quo arrangement, I massage her back for a bit. Her skin is a deep, Mediterranean brown, loose and weathered from 50 years of much sun and little sunscreen. My own aching muscles find some relief in a massage conga line, with a dozen marchers seated one behind the other, each rubbing the back and shoulders of the one in front, the person in front moving to the back of the line after a few minutes.

Later, there's work to be done. In the months leading up to the March, I had drafted a governance structure in which staff would manage big-picture decisions relevant to route, rallies, media, finances, and data. The day-to-day operations of the March would be overseen by an elected council of marchers, with much of the decision-making delegated to working committees exercising a great deal of autonomy.

We have yet to elect a March Council as it seems wise to wait until marchers get to know each other better. But the immediate, essential tasks — cooking, cleaning, setting up and breaking down camp, loading and unloading the gear truck, emptying the commodes — are basic functions that can't wait on an election.

Lala suggests a meeting to sort it out. By the end of a fairly efficient process, work assignments have been established and marchers agree to let Lala delegate duties on a weekly, rotating basis.

The next day's march is longer, weaving through residential neighborhoods with trees and plants so foreign to me they might as well be from the other side of the world. I walk for part of the day with Martin Hippie, who plans to be with us through Redlands. Gifted with an impressive stride and signature determination, Martin was the only marcher to walk every step of the way on the 1986 Peace March.

Martin is an interesting combination of prophet and curmudgeon, offering an authoritative litany of concerns and criticism. He predicts with confidence that we won't make it through the Mojave Desert and tells me candidly that I'm either brilliant or stupid to bring people together from across the country without having all the important details figured out. I listen, knowing he has no comprehension of how much work and planning have already gone into the March and how many obstacles we've already met and overcome. I compliment him several times for having gone the distance on the Peace March, and inwardly express gratitude that I only have to deal with his negativity for one week.

The March passes beyond quiet residential streets into an industrial zone of sacrificial land given over to the broken remnants of America's national religion — consumerism. The sidewalks are often narrow and crumbling, providing virtually no protective barrier against oncoming trucks and cars. This is our first experience walking in dangerous traffic. It's nerve-racking and will become a frequent feature of our routine for the next eight months.

Among a long list of poor choices was the decision to continue broadcasting my live talk show four times each week using my cell phone. I didn't want to cancel the show, as I anticipated it being my primary source of income after the March. Somehow, I needed to keep it going. Besides, I thought my audience would find reports from the March instructive.

Today is my first attempt to broadcast the program from the road. With less than an hour before airtime, I ponder my options for finding a quiet place. I spot a high dam running parallel to our route. I climb the dam and arrive at the top to discover that traffic noise has been replaced by a howling wind. With minutes to spare before the start of the program, I tuck under some rocks behind a stone structure and successfully conduct the broadcast.

This is an absurd, almost comical way to host a radio program. Yet the first broadcast works, and this hit-or-miss search for a makeshift "studio" and adequate phone service would become a predictable feature in the unpredictable life of a wandering talk show host.

As soon as the show ends, I hurry to camp, as I've received my first work assignment from Lala. She's tonight's chef and I'm one of her food prep assistants. The operation takes longer than expected. With the meal running late and dusk settling in, hungry and mildly perturbed marchers approach the kitchen area, walking toward us like zombies. I'm impressed with their sense of humor, despite the travail of the first three days.

That night, I join an impromptu party of a dozen marchers crammed into one large tent. After good conversation enlivened with cheap wine, I retire to my tent to enjoy the best night's sleep yet since arriving in California a week ago.

In the morning, the breakfast crew fails to grasp the necessity of adequate calories for a physically active tribe of hungry nomads. Our rations are one slice of bacon and half an egg. Fortunately, there's plenty of bread, peanut butter, and jelly. I make myself two thick sandwiches and admit that, if I'd been able to raise more money, I would have hired a professional chef.

As we continue our long haul through Los Angeles' sprawling suburbs, snow-covered peaks rise to our left. The path takes us along Route 66, with abundant opportunities to stop for conversation with locals and plenty of food options to make up for the meager breakfast.

Another part of my now daily ritual is the many hours spent receiving and making phone calls and answering email. This slows me down considerably, and I'm again the last marcher to arrive at our destination in Claremont. Here, marchers spend the night with hosts at home stays organized by a coalition of local congregations.

That evening, marchers and hosts assemble at one of the churches that organized our visit. The program includes comments from church members working to address climate change. Each marcher shares their reason for joining the March. A donation bucket is passed and our hosts contribute $500 toward our expenses.

On the altar is a beautiful grand piano. I offer to perform a selection for the audience, and play Chopin Nocturne Opus 9 #2. When I return to my seat, I notice Lala looking at me with tears in her eyes. She later tells me that nocturne is one of her favorite pieces of music.

Already, after only a week of life together, I feel so connected with my new community. I appreciate Lala's sensitive response to my piano playing. I'm impressed with her leadership ability and willingness to step forward and organize the work teams. There are others on the March, too, who have worked tirelessly to plug the gaps and assure our success thus far.

As I head to my home stay, I think about Grace and wonder what she's doing tonight. Despite a few bumps over the past year, our bond had grown stronger during the final three weeks before my departure. Yet over the past week, she's been less communicative and strangely distant. It makes me uneasy. But sleep comes quickly, and I'm grateful for the small comforts of modernity that one normally takes for granted.

Rally at Wilmington Waterfront Park before the start of the March.
(Photo by Jonathan Lee)

The March begins. After two miles, the rain starts to fall and continues for the rest of the 19.4-mile day in Los Angeles, California. (Photo by Jonathan Lee)

Marchers circle up before braving traffic through Los Angeles' suburbs, which took six days to cross.

Shira Wohlberg, Ed Fallon, Ethan Phillips, and Luke Davis walk through a forest of wind turbines east of Cabazon, California.

Jeffrey Czerwiec stands in the wash we follow to bypass Highway 62 heading toward Joshua Tree, California.

Jimmy Betts.

Shari Hrdina.

A MEMOIR FROM THE GREAT MARCH FOR CLIMATE ACTION

Our camp near Twentynine Palms, California.

Kim Foley and Ed ham it up.

A MEMOIR FROM THE GREAT MARCH FOR CLIMATE ACTION

Peter Clay exits the Glass Outhouse, at The Glass Outhouse Art Gallery near Twentynine Palms, California.

A man drowning in sand — one of hundreds of art installations at The Glass Outhouse Art Gallery, near Twentynine Palms, California.

Jimmy and Ed celebrate St. Patrick's Day in the Mojave Desert. David Zahrt is in the background.

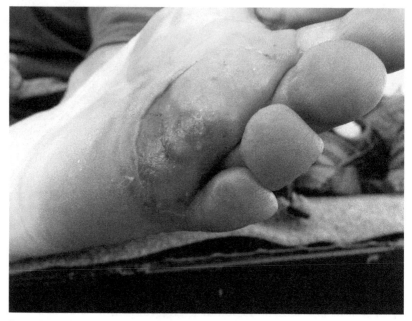

Steve Martin's blistered right foot. The left one wasn't much better.

Jane Kendall (left) and Mary Addams (right) walk with Steve as he navigates the Mojave Desert on crutches.

Marchers gather for a photo after completing the trek across the Mojave Desert. Top row: Shira Wohlberg, Ethan Phillips, Sarah Spain, Izzy Mogelgaard, Mack Wilkins, Luke Davis, Ben Bushwick, John Abbe, Jeffrey Czerwiec. Middle row: Liz Lafferty, Debaura James, Michael Zambrano, Bob Cook, Lala Palazzolo, Jane Kendall, Pablo Howard, Erica Cheshire, Ken Snyder, Kim Foley. Bottom row: Kelsey Erickson, John Jorgensen, Sean Glenn, Marie Davis, Brandon Cheshire, Miriam Kashia, Mary Addams, Jimmy Betts, David Zahrt, and Lee Stewart. (Photo by Ed Fallon)

The Climate Justice Gypsy Band continues to play after crossing the Colorado River into Parker, Arizona.

Native children from the Colorado River Indian Tribes in Parker, Arizona, pose with John J, Ken, Mack, Sean, and Sarah.

Steve and Shira.

Shira sorts through a pile of elk bones.

Ben and John J enjoy supper as the sun sets in the Sonoran Desert of Arizona.

Mark Creekwater with Lala and Ed.

Sean and Mack.

Sarah preparing to leave some of her father's ashes on a
mountainside along the road leading to Payson, Arizona.

A sign greeting marchers at one of our Arizona campsites reads, "Climate change is a hoax! You are all brainwashed by big government." Below that, someone wrote, "and the earth is flat, too!!"

Our solar cookers in front of the kitchen truck (left) and gear truck (right).

Kim, Kathe, and John J cleaning the commodes and preparing to bury the "humanure."

The satchel given to Ed by Sumitra Kulkarni, granddaughter of Mahatma Gandhi, begins to fall apart in Arizona.

Ed and Sarah working as Shira looks on.

Ed, Jimmy, and Sarah with one of our Zuni hosts.

Ed in front of his tent in New Mexico.

The inside of Ed's tent.

Ed snaps a photo of a rattlesnake three feet away.

David Harrington, a local farmer, leads marchers for three days along the obscure track of the original Route 66.

Berenice Tompkins, Ethan Phillips, and Ed wash dishes.

A marcher succumbs to the harsh conditions of the high desert. (Photo by Sarah Spain)

A Native man drums for marchers during a visit to our camp with his family. Eleven Indigenous nations welcomed us across their land during our time in New Mexico.

Danny Lyon and Ed swing in front of Danny's home in Bernalillo, New Mexico. (Photo by Nancy Lyon)

Judy Todd and Ben Bushwick take a break under a bridge near Albuquerque, New Mexico. (Photo by Danny Lyon)

Ed poses with the statue of the pilgrim at El Santuario de Chimayo, New Mexico.

A MEMOIR FROM THE GREAT MARCH FOR CLIMATE ACTION

Steve, Ed, Tony Pisano, Shira, and Trish Booth. Trish and her husband, Leonardo, made breakfast for us at their home in Truchas, New Mexico. (Photo by Leonardo Booth)

Kat Haber, Ed, Ahni Rocheleau, Lala, and Kathe receive foot massages at New Buffalo Center, Arroyo Hondo, New Mexico. Tony and Shira are in the background.

CHAPTER 8

WIND AND SAND

"The desert takes our dreams away from us, and they don't always return."

— PAULO COELHO, **THE ALCHEMIST**

At 19 miles, today's hike is challenging. Many of us are still out on the route well after dark, walking along a road unsafe even in daylight. We warn oncoming motorists of our presence as best we can by waving our cell phones, hoping their collective glow will be visible. I add to my growing list of mistakes forgetting to bring a headlamp, and remind myself to buy one in Redlands.

Tonight's campsite is dismal: a small, cramped parking lot adjacent to a mosque. In addition to the discomforts of a concrete surface and relentless traffic, neighboring roosters crow loudly most of the night. My earplugs prove worthless against their incessant racket.

Functioning on just a few hours sleep, another long hike is made bearable by the promise of our first day off since the start of the March. During the last couple of miles, we see more and more orange trees. I pick two oranges from branches draping their bounty over the sidewalk. I tell myself that the owner wouldn't care that a hungry walker picked fruit that had strayed into the public domain. I'm certain my conscience is in the clear, but my mind insists on an internal analysis of the nuances of honesty, compelling me to justify the moral rightness of dining on this perhaps forbidden fruit.

Once again, I'm the last marcher to arrive at our overnight stop at Redlands United Church of Christ. At least it's still daylight. I'm tired but content, my cheeks and chin dripping with the warm sweetness of fresh orange juice — ethically acquired, I surmise.

A day without marching! People sleep in, wash clothes, and stroll around the beautiful church gardens abundant with lemon, lime, and orange trees. A mischievous marcher convinces another that a lime is simply a baby lemon that will, in good time, grow up and turn yellow.

It's a day off from marching but not a day off from work. Several marchers tackle the challenge of installing shelves in our gear truck. I visit the town center and buy a headlamp, then practice for a full hour on the church's baby grand piano. It's such a joy to play again. Several marchers linger, including Lala, who sits through the entire "concert."

Most of my day is spent catching up on March work — raising money, contacting the media, and figuring out how to navigate the Mojave Desert. Shari has been on the March for the first two weeks, although today is the first day she and I have been able to work together.

Shari's not the least bit excited about camping, marching, bad weather, or pollen. Meals are problematic as she doesn't eat garlic, onions, peppers, or anything spicy. There's a certain stoicism about her, amplified by her extremely long hair and a presence that suggests "New-Age Amish." Despite the physical discomforts of the March, Shari soldiers through the obstacles and provides a stability and serenity to our community that will be missed when she returns to Des Moines next week.

Refreshed after our day off, marchers set out in high spirits as the suburbs of LA end abruptly. The landscape shifts quickly from lush to dry, the dryness interrupted frequently by sprawling oases of orange groves.

We pass a herd of zebra. I presume they're purely for recreational purposes, though the omnivores among us joke about the health benefits of free-range, grass-fed zebra steaks.

We pass through Beaumont and arrive at Cabazon. The landscape grows drier and ever more rugged. Orange groves give way to wind turbines. Where there was soil, there is now mostly sand and rock. Sandwiched between Interstate 10 and busy train tracks, our cramped campsite is an unpaved parking lot that doubles as storage for young, potted palm trees. The long palm fronds cut easily if one is careless enough to touch them.

As I prepare to set up my tent, a police officer pulls up. "Can I speak to the person in charge?" he asks, his large frame slightly imposing.

"I can help you," I offer, presuming we're in some kind of trouble. Perhaps the LA County Health Department has finally caught up with us and our truck is about to be whisked away to imprisonment in a commissary.

"Do you know much about Cabazon?" the officer enquires.

"No, not a thing," I reply. "We just got here, and I don't think any of us have ever visited before."

"Well, there's a lot of down-and-out types in Cabazon," he explains. "We've got our share of alcoholics, and some will probably try to make themselves at home in your camp, if you're not careful."

His warning proves accurate. During dinner, several drunk men show up. Mack Wilkins, one of our youngest marchers, invites them to join the supper line. The task of bad cop is left to me and I politely but firmly ask them to leave. I catch flack from Mack, who feels that everyone, even the chronically inebriated, should be welcome in our camp.

That night, as I try unsuccessfully to sleep, sandwiched in between the intense roar of Interstate 10 and trains that whistle and rumble every 15 to 30 minutes just a stone's throw away, I consider that perhaps the only way a person can sleep in Cabazon is with one's brain numbed by excessive quantities of alcohol.

The next day we walk through the heart of the region's massive wind farms, up a mountain and literally right under the blades of giant turbines that even the mightiest Don Quixote wouldn't dare joust. I'm well-accustomed to wind turbines in Iowa, but they're rarely situated where one can walk up to them, touch them, and get the full sense of their massiveness. The sand, the mountains, the constant wind, the close proximity of the turbines — all make for a futuristic, almost other-worldly experience. With the sun beating us down, we find moments of relief standing or sitting in the shade of the towers.

The turbines go on as far as the eye can see — and the eye can see quite a ways from our vantage on the mountain we cross to get to today's campsite. It's encouraging to see such progress with renewable energy, a positive impression that will be off-set by the frequent trains carrying coal and oil that will pass us day and night in Colorado and Nebraska.

The next day's walk is grueling. We lean hard into a steady 35-mile-an-hour headwind. My straw hat — essential for keeping the sun off my head and face — becomes a liability, catching the wind and straining my neck muscles. I watch Miriam Kashia, age 71, walk into the gale at a 45-degree angle.

Miriam is a retired psychologist and a veteran of back-country adventures. Months prior, she'd prepared a video for marchers on what to expect and how to pack. Yet already we've confronted conditions, like today's fierce wind, that none of us, not even Miriam, were prepared for.

We arrive at a dangerous stretch of Highway 62. Our original plan had been to stick to the road, stay close together, and hope for the best. But Mack did some scouting and found an alternate route, one that winds through a wash and up over rocks to flatter ground beyond. The terrain is challenging, but we're spared the dangerous traffic and blessed with a glimpse of several big horn sheep.

Once over the pass, it's again safe to walk on the highway — as safe as walking on a highway can be. I stop twice to attend to March work and fall two to three hours behind. I land in Joshua Tree National Park, our camp for the night, well after dark and dinner. The chef has kindly saved me some food, cold but nonetheless satisfying. In the light of a half moon, I'm in awe of the fantastic shapes of the Joshua trees, limbs bent and twisted, like scraggly arms reaching for a distant star in the chill desert sky.

I eat quickly and slip into my tent. The cold makes for a restless night's sleep and my phone alarm wakes me too soon. Aching limbs protest strongly as my mind orders them out of the tent and into the cold morning darkness. An hour and a half later, after breakfast and breaking down camp, the eastern sky is still pitch black as we set out. I wave goodbye to a pair of saluting Joshua trees, regretting that I never did get to meet them in daylight.

Today we walk along Highway 62 to Joshua Tree Retreat Center, west of Twentynine Palms, where we enjoy a day off before attempting the 125-mile, nine-day crossing of the Mojave Desert — a crossing that many, not just Martin Hippie, predict we won't make.

For two days, One Earth Village is a bustling hub of activity. We stock up on food and water for the desert crossing. Marchers paint the gear truck. We swim, relax, do laundry, and celebrate the wedding of Ki and her fiancée, Taylor Miller.

It's a wonderfully simple wedding, appropriate to the desert landscape. Jimmy officiates and I play classical guitar as marchers form a circle around the couple, barefoot in the warm desert sand.

Later, marchers are invited to another celebration. In the hills west of town lives Bobby Furst, an artist who's astonishingly unique in a profession where uniqueness is normal, almost required. Bobby throws a magical party that night in appreciation of our efforts. By the time I arrive the food is long gone, gobbled down by other guests and marchers with insatiable appetites.

But there is ample sustenance for the spirit. Bobby's home and yard are crammed with art made almost entirely from the flotsam and jetsam of American consumerism — what most of us would call "junk." The creation that moves me the most is the word "Peace" written in bullets, accompanied by a book listing the names of those who've died in recent wars. Bobby says this work speaks to the senselessness of war and the squandering of lives, time, and money.

Inside, a band plays while marchers and other guests drink and dance. The evening is primal, Bacchanalian, a hearty toast to the human spirit freed of the constraints of convention. Some marchers lose themselves in the music, alcohol, and seduction of crisp, moonlit desert air. But I'm exhausted, and after a couple hours catch the first shuttle back to camp.

A MEMOIR FROM THE GREAT MARCH FOR CLIMATE ACTION

The next day, just beyond Twentynine Palms on the edge of the great Mojave Desert, we camp at one of the most unusual places to welcome us during the March. Next to their modest home, Laurel and Frank operate The Glass Outhouse Art Gallery. It's a bizarre place, as artistically odd and alluring as Bobby Furst's home. It has the feel of the last outpost at the edge of the world — a final safe, welcoming domicile before adventurers and madmen plunge into a realm where human life treads on a tightrope of what is biologically possible.

The art is clever, funny, recycled, and everywhere. There is a life-size image of a man buried up to his torso in sand, clawing desperately at the air, trying to free himself from his desert prison. A caged coyote stands calmly next to a lamb. Nearby is a "pond" of sand marked with a sign reading "Frog Parking Only — All Others Will Be Toad." I joke with Bob Cook that, given his sleep apnea and propensity to snore like a bull frog, maybe this can be his new home after the March.

And of course, there's an outhouse made of glass. Looking in, one sees nothing of the interior. But once inside, a panoramic view of the outside world is on display. I enter the outhouse and sit for a minute, trying to understand why I find it uncomfortable, why I would never dare defecate in public.

To pee? Sure. But to shit? Not a chance. Is this inhibition merely cultural, or is there a biological imperative? Perhaps it's ingrained in humans to shit privately because we are so vulnerable at that particular moment. Was defecating out of sight of predators originally an act of self-defense? I sit in the outhouse, imagining a lion or tiger coming upon a squatting humanoid. Trapped in mid-dump, such prey would have made an easy catch for a large carnivore.

I convince myself that, whether cultural or biological, the ingrained prohibition against shitting where others can see you, or where it appears they can see you, is silly. Nonetheless, I exit the Glass Outhouse never having dropped trou and I find my way to the now familiar comfort and privacy of the Eco Commodes.

Our night at the Glass Outhouse is not without conflict. A problem visitor has tracked us these past few days — a man named

Adam, who presents himself as a medic, eager to assist the March. There's an unsettling oddness about him, and when I interrogate him while we walk, it becomes clear that his long list of qualifications is an elaborate fabrication.

Adam's transformation from odd to threatening doesn't take long. Several women complain to me about him. Lala is outraged when he starts kissing and caressing her feet. Adam then finds his way into Kim's tent, uninvited and unannounced.

I'm furious. I confront Adam directly. In strong words, I tell him he must leave immediately. He refuses. I insist, get in his face, and he leaves. Some marchers grumble. Mack is uncomfortable with how I handled the situation, but most seem relieved that Adam is gone.

I settle in for the night. Staring out the door of my tent, I'm mesmerized by the decorative lights on Laurel and Frank's home, shimmering as they blend with the moonlight to illuminate Laurel's art, alighting on clay and rock and glass like displaced prairie fireflies dancing above the desert sand. I feel a tinge of home-sickness, and wonder if fireflies will still light the summer night sky when we walk through Iowa in August. I miss the seedlings I normally tend this time of year. I miss my hens, friends, cat, piano, and home. I miss Grace, and as thoughts of her invite my mind to slumber I'm jerked back to full consciousness by the reminder that tomorrow begins what could be the most daunting stretch of our 3,000-mile journey.

We passed a sign earlier today that warned of no services for the next 100 miles. Shortly after, as if sent by a higher power to hammer home the severity of that warning, a man in his 60s driving a beat-up Chevy pulls alongside us. He rolls down his window, stares for a few seconds and asks "What the hell are you doing?" his raspy, caustic voice sounding more annoyed than curious.

"We're walking across the country to get people to understand the urgency of the climate crisis," I respond.

"Aw, you don't believe that climate-change bullshit, do you?" the man mocks, certain of his own unshakable scientific literacy. He pauses for several seconds, glancing up the road toward the vast expanse of the Mojave, toward a landscape that, from my point of view, appears as forbidding and foreign as the moon.

"You aren't planning to walk across *that*, are you?"

"Well, it's the Great *March* for Climate Action, not the Great *Car Ride* for Climate Action," I dead pan, abandoning the prospect of converting my inquisitor, and embracing this opportunity to let my inner smart-ass frolic.

"You're fucking out of your mind!" he yells. "Have you been out in the sun too long?"

"Yes," is my simple and honest response, as the man drives away muttering more expletives and something about dying in the desert.

"Nice guy," I mumble.

We laugh off his prognostications of our impending demise. Yet stuck in my mind is Martin Hippie's more thoughtful, well-intended warning that we won't make it across the Mojave because we don't have our food and water figured out.

But I've seen how the March community has pulled together these first two weeks, tackling numerous logistical challenges that our staff and I admittedly should have had in place prior to the start of the March. Between the collective brain power and brawn of this determined group of climate warriors, I'm not worried about food and water.

A different concern troubles me, and troubles me deeply — a concern that indeed could prevent us from making it across the desert.

We have no place to camp.

Despite my efforts to reason with BLM officials, they refuse to grant permission for us to pitch our tents on BLM land anywhere in the Mojave. And nearly *everything* in the Mojave is BLM land.

I crawl deeper into my sleeping bag for one final night of comparatively civilized camping. Tomorrow begins a nine-day trek across a parched, barren landscape hostile to most life on Earth. I consider the very real possibility that BLM officials will order us to leave the desert or arrest us if we refuse, wondering if Martin Hippie and the Chevy-driving curmudgeon might be vindicated after all.

CHAPTER 9

MOONWALK

"The Mojave is a big desert and a frightening one. It's as though nature tested a man for endurance and constancy to prove whether he was good enough to get to California."

— JOHN STEINBECK

One of our marchers, John Jorgensen, is a biology teacher from Tucson, Arizona. John is intimately familiar with the challenges of life in dry country and regularly shares with us all manner of desert wisdom. Before putting on our shoes in the morning we check them for scorpions. We learn the names of cacti. John warns that the desert will rattle your teeth at night after it slow-bakes your flesh during the day.

John knows the desert. I know politics, and I've been struggling with what seems to me an essentially political problem: the BLM's refusal to let us camp in the Mojave.

I turn for help to US Secretary of Agriculture, Tom Vilsack. Tom and I were both elected to the Iowa Statehouse in 1992 and got along reasonably well. But after Tom became Governor in 1998, our rapport soured and just kept getting worse.

We were at odds over hog confinements, which I felt needed greater regulation. We disagreed over the "Official English" bill, which I fought hard against and he signed. We were at odds over his center-piece legislation — the Iowa Values Fund, which I saw as corporate welfare. We strongly disagreed on eminent domain, which I felt should not be allowed for private development.

We were also at odds over eliminating the sales tax on utility bills, which I strongly supported and Tom opposed. In 2001, the double gut-punch of a really cold winter and exceptionally high fuel prices saw utility costs soar. Low-income people with drafty old homes were hit particularly hard. The Iowa House and Senate responded with a bipartisan bill to eliminate the sales tax on utility bills. Governor Vilsack threatened a veto.

I realized a high-visibility campaign was needed. I orchestrated several events to pressure Vilsack to sign the bill. The most colorful brought poor families to the steps of the Statehouse where we lit a bonfire of old chairs to symbolize the choice between heat and other essentials that low-income homeowners faced that winter.

It's impossible to know how much influence the bonfire had on Vilsack's decision, but the next day, after a private meeting with House and Senate Democrats in which he angrily berated those of us who supported the repeal, Vilsack signed the bill into law.

In 2007, Tom and I were both done with our time in state government. I'd lost my primary for governor and Tom had dropped out of the race for president. I suggested we get together to try and move beyond our differences. A string of "coffees" mended fences, and we agreed to collaborate on pushing campaign finance reform forward.

As I weigh the March's options for crossing the Mojave Desert without a permit to camp, including the very real possibility that the March might not be able to continue, my time spent mending fences comes in handy. Tom tells me he'll do what he can to help, and I wait to see what will happen.

In the meantime, we keep marching, working our way through an ecosystem that is uniquely minimalist, stark, and barren. Yet even here we see occasional beautiful flowers, compliments of a rare recent rainfall, made more spectacular by their singular isolation on a pallet of sand and rock.

The Mojave's rough terrain provides few functional campsites along Highway 62. Most are littered with broken glass and shot-gun shells. It strikes me as absurd that the BLM does little to prevent or clean up this desecration of the natural landscape while denying a permit to a group of environmentalists. We pitch camp amidst the shattered reminders of America's obsession with firearms, though I can't shake the expectation that, at any moment, officials will pull in to our camp and order us out of the desert or into paddy wagons.

Three days into our desert crossing with no visit from BLM officials, our biggest threats are heat and cold. I meticulously embrace John's desert advice on all accounts and presume that my straw hat, jeans, and liberal applications of sunscreen will keep me safe during our daily 12- to 17-mile walks under the scorching desert sun. Yet over the past week my lips have become dry and cracked. The problem only gets worse in the Mojave. Both lips are now bleeding. To make matters worse, one lip develops a cold sore. The other, not to be outdone, soon follows with its own stylish blister.

As a person who spends a lot of time outside, I'm used to dealing with extreme temperatures — but not both extremes in one 24-hour period. The nights are so cold that even my new sleeping bag is ineffective.

The best remedy against the cold proves to be a tent partner. More and more marchers pair up, some as couples, others simply for warmth, companionship, and to expedite the setting-up and breaking-down of camp.

I offer to share my tent with Shira Wohlberg, whose tent broke even before the March started. Shira had suffered badly on the first day of the March, irritating an old knee injury from her days as a gymnast. She'd intended to walk the entire distance only to be sidelined immediately. I felt bad for her.

But Shira was tough and determined. She recovered quickly and soon became one of our fastest marchers. She and I enjoy walking together until that point in the day when, inevitably, I stop to attend to March business that I can't accomplish while walking.

Shira is a naturally affectionate person — a trait that is magnified in One Earth Village, where hugs and physical contact are prevalent far beyond what is typical in conventional society. Shira will cuddle up with another marcher, male or female, for no apparent reason.

Sometimes during meetings, she sits on my lap. Sometimes, she sits on someone else's lap. There's no awkwardness about it, and I wonder why the "real" world can't be more like this. If a marcher looking for affection saunters over to another, that marcher will be held, have their neck and shoulders rubbed, their hair brushed. In "civilized" settings, only cats, dogs, and babies get away with such behavior. We are a society that fears intimacy even as we're obsessed with sex.

Yet, revealing my naiveté in matters of love, I should have know that Grace would be upset with me having a female tent partner. She's upset, but not angry, and over the next few weeks becomes less aloof than she's been. She starts calling more often and sends photos of herself, something that almost never happened before. If anything, the tent-mate development brings Grace closer and mends that inexplicable distance I'd detected during the first two weeks of the March.

Sharing my tent with Shira elicits a response from one other person. As I'm trying to anchor a corner of the tent in a surface that is basically rock, Lala walks over with an inquisitive look on her face.

"Are you and Shira sharing a tent?" she asks. I tell her we are. She mutters something perfunctory, along the lines of, "I see, that's nice." The moment is awkward, and as I try to read between the lines, I sense confusion and disappointment. I have the strong impression that Lala had hoped to be my tent mate, and for reasons beyond comfort and survival.

On Saint Patrick's Day, under a clear sky and brilliant full moon, we set out from camp at 2:00 a.m. If one walk could be described as perfect, this was it. Visibility across the moonlit, moonlike landscape is ideal, without the torment of daylight blinding and burning us. Traffic is almost non-existent. We walk together, interspersing our conversation with periods of comfortable silence, reveling in the quiet desert night.

Mack finds a plastic plate in the ditch which he and I use as a frisbee. We claim the full width of the road as our fairway, hurling the plate-turned-frisbee back and forth as we walk eastward.

We arrive at our campsite just as the sun rises. Jimmy and I climb a hill on the edge of camp and play a few Irish tunes to welcome the dawn. It's a unique celebration of my ancestral holiday — no green, no parade, no beer. Instead, a night-walk in the desert, an impromptu sunrise jam session, then a long nap in the shade of the gear truck as the day moves quickly from cold to hot.

On the sixth day of our desert crossing, a car slows as it pulls alongside us. A man wearing a BLM uniform rolls down the window and asks, "Is one of you Ed Fallon?"

"Yeah, that's me," I say, thinking this might be the moment I've feared, when federal agents throw us out of the desert.

The BLM official pulls over, gets out of the car and walks over to me. "Here's a letter I'd like you to sign," he says. "It gives you permission to camp anywhere you want on BLM land in California. We're working with Arizona and New Mexico to make sure you have no trouble there either."

I thank the officer as I sign two copies of the letter, stuffing one into my satchel. As soon as we have an Internet connection, I thank Tom Vilsack profusely. In terms of our trek across the Mojave, it's a landmark moment, but not without a tinge of bitterness. It's sad that one needs a personal connection to someone with clout in order to get the massive wheels of government to function rationally.

Two days later, we arrive at the first manmade structure we've seen in over a week at an intersection called Vidal Junction. Marchers swarm the lone convenience store, tossing aside dietary scruples to snag sugary, high-calorie treats. I practically inhale two ice cream sandwiches and contemplate a third, but resist.

Tomorrow, we cross the Colorado River into Parker, Arizona. Even if the March goes no further, what a tremendous accomplishment to have survived the Mojave Desert! I'm in awe at how everyone has pulled together. Our vehicles only had to leave the desert once to refill the water tanks. We desperately need bathing and clean clothes, and many marchers experience some discomfort from blisters. My feet remain blister-free, as my body prefers to cluster its blisters on the lips.

The Mojave claims one casualty: Steve Martin, who on day seven of our desert crossing developed the worst case of blisters I'd ever seen. Steve finishes the day crippled and hobbling. I comment that

I've seen roadkill that looked better than the bottoms of his feet. Of all marchers, Steve is the most determined to walk every step of the way. Today, we doubt that will be possible.

Steve finds some relief as Jimmy applies his knowledge of Eastern medicine and Kim her skills in Western medicine. Steve's condition is stabilized, and over the next two days he walks the last 33 miles to Parker on crutches.

Our determined community of marchers has cooperated in so many ways, sacrificing and adapting to meet one seemingly insurmountable obstacle after another. I feel as though I'm part of the most successful, functional, loving family in America.

As we prepare to leave California, I never could have imagined how precipitously this beautiful spirit of cooperation was about to disintegrate.

CHAPTER 10
THE HONEYMOON IS OVER

"Always remember that others may hate you but those who hate you don't win unless you hate them. And then you destroy yourself."

— RICHARD NIXON

On the morning of March 23, thirty of us assemble on the California side of the Colorado River, feeling every bit like conquering heroes. Police officers escort us across the bridge. For a short stint the March again becomes a parade — albeit a much smaller parade than the 1,500 climate walkers who clogged the streets of Los Angeles three weeks earlier.

For me the moment is exuberant. The March has successfully navigated the Mojave Desert, completing one-tenth of its journey. Despite the back trouble that laid me out in February, I have so far been able to march every step of the way.

Triumphantly, we cross the bridge into Arizona bellowing six verses of *This Land Is Your Land*. I sing so loudly my vocal cords strain. At the bridge's midpoint it seems our voices can be heard on both sides of the river. Several of us play guitar, Jimmy the fiddle, and Lala and Marie percussion. We are a dirty but jubilant band of climate warriors, survivors of the Mojave Desert.

A line of drivers piles up behind us, enhancing the parade-like atmosphere of our crossing. I convince myself that the drivers tailing us at 3 miles an hour are delighted we're here, happy to delay the progress of their day to bask with us in this great moment of accomplishment on this singularly important mission. A few drivers share hand gestures that suggest otherwise. My exuberance remains undaunted.

Our Arizona coordinator, Erica Cheshire, has secured a wonderful campsite in Parker: a shaded strip of grass on the Colorado River Indian Tribes reservation. We'll spend two nights here, enjoying our first day off since the Glass Outhouse. A handful of tribal children greet us. We outfit each of them with a Climate March t-shirt and they're eager to help set up our tents. I attract the largest group of assistants and John Jorgensen jokes that this will be the first time my tent is set up properly.

A few marchers opt to stay at a hotel, allowing the rest of us to enjoy a shower and the hotel's swimming pool. Parker is a small town of 3,000, but given the amenities-starved enthusiasm with which marchers invade its restaurants, bars, laundromats, and movie theaters, Parker might as well be New York, Paris, or London.

Our day off does wonders for Steve. His feet recover enough that he's able to resume walking without crutches, although his armpits will remain sore for weeks.

After our day off, we head into the Sonoran Desert, which is lush compared to the Mojave and abundant in plant and animal life. In addition to the wealth of flora and fauna, our route through the Sonoran is abundant in traffic. Every fourth vehicle is a pickup truck, seemingly driven by a guy powered by excessive testosterone and gifted with little common sense. When he and his ilk barrel past us and refuse to move over, our walking conditions are unnerving and often dangerous.

When topography allows, we escape the road and follow washes or fence lines that parallel the highway. These prove safer — but the rocks, ravines, loose sand, and fences are difficult to navigate. Mostly, we're stuck on highways. We experiment with various strategies to keep from being mowed down, including marching tightly together with one of us, often me, 100 yards ahead feverishly waving a red flag at approaching traffic. This proves effective at getting drivers to slow down, often appreciably.

But the fact that we're marching across Arizona at all is a small miracle. Two weeks earlier, an Arizona Department of Transportation official informed us in no uncertain terms that we were forbidden to walk on state highways unless we purchased $5 million worth of insurance, hired two vehicles with flashing lights to accompany us, and paid the nightly hotel costs for the drivers of those vehicles for an entire month.

Effectively, the official told us we weren't going to march across Arizona.

Yet we had no choice. We entered Arizona in the middle of the state with no possibility of detouring north or south. This was a challenge as serious as our BLM problem, and since it involved state law, I didn't have Tom Vilsack to help me out.

A week ago, I called the DOT official who had issued us the ultimatum. I tried to break the ice and jokingly offered to cut costs by feeding and housing the drivers in tents in our camp. "They'll love the couscous, tofu, and fresh vegetables," I assured him. My attempt at humor failed. The call went poorly and the dour official refused to budge.

A couple days later I managed to reach his supervisor. That conversation also ended at a firm impasse.

My only option was to pull rank on the two mid-level bureaucrats. I called the director of the DOT a couple days before our scheduled arrival in Arizona. I introduced myself as a former Iowa legislator, familiar with laws governing equal and fair access to public thoroughfares. In the process of our conversation, it became clear that the problem was the word "event." I explained that there'd been a misunderstanding. "We're not an event," I told him. "We're simply thirty people out walking for a good cause."

I shared a few examples with the director. What if three guys go for a run on a state highway? Do they need insurance and have to be accompanied by vehicles? What about six people out for a walk? Or a dozen people riding bicycles?

The director agreed that these assemblages were legal and required no special permits. "Well, that's us," I assured him. "We just have a fancy name, an important purpose, and a long way to walk."

I told the director I appreciated his commitment to our safety and the safety of drivers. "That's a commitment we fully share," I said. "We're always deferential to cars and trucks that pass us and we're used to walking in traffic. We don't want to get hurt, and we certainly don't want to cause any accidents."

My argument proved persuasive. The director conceded that we were within our legal right to walk Arizona's highways without a permit, and the March scored its third victory in the battle between common sense and bureaucracy.

Halfway from Parker to Phoenix we have another rest day. Usually, our day off is in a big city or mid-sized town. Today, it's a sandy parking lot next to Western Trails Ranch restaurant. Marchers patronize the restaurant extensively, and the owner enjoys what is perhaps her busiest two days of the year. For the omnivores in our group, the cook prepares a steak dinner, which I find unusually delicious. We pack away hefty portions of food and drink, play pool, and dance, with Shira and Lala performing some impressive, improvisatory choreography.

The great fraternity we've developed and maintained across one-and-a-half deserts seems destined to continue.

My cat's name is Mika. She had an interesting and tragic start in life. While still in the womb, her mother was hit by a car. The mother lived, but all the kittens except Mika were born dead.

I took Mika in when she was ten days old, so small I could carry her in the palm of my hand. Later, I trained her to ride on my shoulder. That became her safe place, her perch during our walks around the block.

Other than riding on my shoulder, Mika was an indoor cat until she was three years old. When venturing outside finally became an option, she would stick her head out the door with great caution, look around timidly, then dart back inside. A world that appeared safe from my shoulder was terrifying at ground level.

Nowadays, she's outside half the time, fulfilling a critical role in purging my garden of mammalian pests. Like many cats, who look so peaceful and harmless resting on one's lap, Mika is the bane of mice, chipmunks, and baby rabbits. In any given year, she'll decimate a few dozen creatures, eating some and decorating the yard and sidewalk with the remains of others. Gratefully, she avoids killing birds, perhaps because she knows I wouldn't approve, or maybe because there are so many juicy mammals to hunt.

One of our marchers, Kelsey Ericson, reminds me of Mika. Kelsey's life has been difficult and recently marred by the tragic death of her boyfriend. She's quiet, walks hard, works hard, but often looks scared, as if hunted by some nameless internal demon.

It's been difficult to get to know Kelsey. She's always been friendly to me, yet over the past few days, I've notice a change. I catch fleeting glances that bely disapproval, even disdain. I sense something is about to break.

When Kelsey finally pounces, the attack is vicious. It comes during a camp meeting. She angrily accuses me as a paid staff person of "profiting" from the March. She wants all staff fired, but for now targets me alone. Kelsey proposes I either not be allowed to march or give up my position as paid staff.

I'm shocked. I knew there would be plenty of obstacles to marching — including injury, illness, weather, altitude, and the challenge of working while walking. But I always presumed marchers would appreciate my commitment to do the same hard work they were asked to do. I never imagined any marcher would be *against* me marching.

Perhaps due to her upbringing, Kelsey seems unable to grasp, for example, that a person hired to write press releases isn't "profiting" any more than a vendor paid to print brochures. Her attitude is even more surprising given the pay structure for staff: all March employees are paid $3,000 per month. That includes me, with 30 years of experience, and Zach Heffernen, our organizer fresh out of college.

Over the next couple weeks, Kelsey's tirade against me gains traction with other marchers. I remind her and others that they signed up to march knowing we had paid staff and that I was the director. I argue that dwelling on this is a drain on our collective time and energy and a huge distraction from the March's focus on climate change.

But the problem gets worse. More marchers jump on the "Ed shouldn't march" bandwagon. Kelsey escalates her campaign to bizarre levels. She doesn't believe that I work while I walk and demands that I fill out a daily timesheet. I refuse. She then says she'll shadow me throughout the day. I tell her that's not shadowing, it's stalking, and I won't tolerate it.

She stalks me anyhow. On a day that promises a beautiful, pastoral walk in traffic-free conditions, every time I stop to take a call, Kelsey stops. She stares at me with a creepy, haunting look, within earshot of my conversation. When I walk, she walks. When I stop, she stops.

I finally explode, and in front of her and other marchers, I announce that, if Kelsey continues, I will call the Arizona Department of Human Services and report her as a stalker.

What bothers me most is that other marchers don't challenge Kelsey, don't defend my right to march, don't speak out about the importance of paid, professional staff to the success of our mission. I tell the March's judicial committee that I'm serious about reporting Kelsey as a stalker. Lala is on the committee and convinces Kelsey that shadowing me is wrong. This helps, but the grumbling against me continues — and spreads to other concerns.

It's been nice having friends from Iowa on the March, especially Kim and Jimmy. Kim and I go back many years. She's my massage therapist. I played piano at her wedding, and she and I walked together many times in preparation for the March. We've bonded over the years, and our bond only grew stronger leading up to the March.

Yet lately, I sense hostility from Kim, too. One day she confronts me. "Ed, you've changed," she says angrily. "You're not the same man I knew back in Des Moines."

I'm puzzled and say, "I haven't changed. You haven't changed either. It's just that none of us have ever lived like this, in such close quarters, doing such physically and emotionally demanding work."

My response only further irritates Kim. "You need to spend more time in camp, Ed. You hardly ever take time to socialize. Marchers need to see you. We need to have you be part of our discussions and conversations."

"I'd love to do that," I respond. "But with my duties as director, talk show host, and marcher, I don't have much time to socialize."

"Well, then maybe Kelsey's right. Maybe you shouldn't march," says Kim.

I'm floored. I don't know how to respond. Angrily, I say, "Well, if you don't see the value in marching, if you no longer understand that that's the focus of what this is about, then it's you who's changed, not me." I storm off.

I turn around and say as I walk away, "You go have fun, Kim," ending the conversation on the worst possible note. "I've gotta raise funds so you and others can keep socializing."

The next day, I apologize to Kim and explain that it's hard to keep your cool in the intense strain of the March. Kim has little to say. Our conversation is brief, and provides only token resolution to the previous day's confrontation.

I face conflicts on other fronts as well. Some marchers, including Miriam, are worried about our financial condition. March finances have always been completely transparent. I've encouraged those who have questions and concerns to contact Shari and study our income and expense data.

Upon learning that we only have enough money in the bank to cover expenses through mid-May, Miriam panics. Our relationship grows tense over concerns about finances. Another marcher, Ken Snyder, is downright angry. Ken confronts me, presuming I'm ignorant of the situation. I calmly tell him I'm well aware of our finances and that having a four-week cushion is more than we've had at other times. I'm not the least bit worried and will continue to work hard to raise money. Other marchers are doing their part to raise $20 for each day they're on the March — $5,000 for those walking coast-to-coast. Twelve marchers have raised more than $5,000. I tell Ken things will be fine, especially if he's willing to do his part. He doesn't and soon leaves the March.

A special meeting is called to grill me. The meeting comes at the end of a hard day's march. I'm exhausted and have had no time to prepare. I answer some questions badly. A few marchers accuse me of mismanaging funds. The meeting resolves nothing, inflames marchers who are worried about money, and gives Kelsey additional ammunition. I retire to my tent and vent for a bit to Shira, but I'm so livid I'm barely able to sleep.

That night, I think of the rattlesnake that joined Shira and me for lunch a few days before. It was hot and we craved a shady spot for our break. We found a stunted tree and climbed its low branches a few feet above the ground where we nibbled on our fruit and sandwiches. My satchel, walking stick, phone, and the peels of an orange lay on the ground in front of me when the rattler approached.

It was the first venomous snake I'd ever seen in the wild. The rattler's reputation for inflicting pain and death elevated the sense of danger as it passed a yard from my dangling feet. For reasons I'll never understand, I abandoned common sense and slowly reached down for my phone. The snake stopped a mere four feet away as I managed to snap a few pictures before it quietly slithered off.

As I lie in my tent, tense and angry from a meeting that felt like a witch trial, I sense that the March has been bitten and venom is now coursing through its veins. Is it too late to recover the wonderful sense of community that prevailed during our first month together? Or has the poison spread so deep that it will only grow stronger and eventually destroy us?

CHAPTER 11
SLEEP

"They will tell you, you can't sleep alone in a strange place. Then they'll tell you, you can't sleep with somebody else."

— BILLY JOEL, "MY LIFE"

The The National Famine Museum in Strokestown in Strokestown, Ireland, is a short distance from my family's farm. I go there often when I'm home visiting relatives. The Famine Museum assaults the mind with a disturbing but important truth. It explains that the Great Hunger of 1845 wasn't caused by a blight infecting the Irish potato crop but by a blight afflicting the heart and mind of Ireland's English oppressors — a blight of conscience and character that enabled Britain to rationalize genocide as reasonable economic policy.

There's a drawing in the Famine Museum that never fails to captivate me. A poor Irish family of seven sleeps on the stone floor of their tiny, cramped home, their pig cozied up next to them. Though destitute and presumably hungry, members of this family at least have the consolation of each other through the night — and the security of knowing their lone source of protein won't disappear while they sleep.

Americans have come to accept as normal the modern sleeping arrangement whereby most men, women, and children are sentenced to their own private bed in their own private room. I remember how desperately my daughter Fionna, at age 7, would plead to sleep with

Kristin and me, sometimes to the point of tears. At the time, I thought she was just "being a baby." Yet she was simply responding to the basic human need to be close to others in sleep.

Sharing my tent with Shira, I rediscover a beautiful part of the human experience that's been missing from my life. Aside from helping each other stay warm, sleeping together brings an internal recognition that, though each of us is alone in the world, it is better to be alone together. In my current dilemma, under attack by marchers on two fronts, Shira's companionship and our conversations before sleep are especially helpful.

We sleep with two bags zipped together and another thrown over the top. Some nights the temperature drops into the 20s. Without a tent partner, even two sleeping bags and long underwear wouldn't keep us warm.

The best part of the day is when we first crawl in and zip up the bag against the cold of the desert night. We're two furnaces and move quickly from cold to warm as we drift off to sleep. Sometimes when I wake in the night, Shira and I are close together, and it's comforting to emerge from a dream — especially an unpleasant one — to find myself near her.

More often as we fall asleep, Shira and I move apart — imperceptibly shifting to different places in the bag. Perhaps our bodies need to cool down. Perhaps we shift because the flesh needs a break, needs space, needs to breathe on its own because the experience of human closeness is too rich, too intense — like a delicious dessert manageable only in small servings.

I smile when I wake and find us side by side. I smile when I find us lying apart. Whether Shira is right next to me or nearby, either feels good, right, and natural. Either is an antidote to the inherent isolation of existence.

Even while sharing a tent with Shira, I miss Grace. The distance that settled in between us during the first few weeks of the March is gone. We talk often. The conversations mostly circle around the day's activities. We steer clear of matters deep and personal, matters best discussed face to face. I share anecdotes from the walk and Grace helps me process some of the difficulties I face with marchers. Grace is dealing with challenges as well, often related to conflicts at work.

I try to help her sort out how best to respond. We don't talk about my tenting arrangement. Perhaps Grace lets it slide because she knows the man Shira is in love with, Tony, will arrive on the March in mid-May.

Tonight as I lie awake, I think of Grace and smile as I remember the night we fell in love.

We were at a dance. Grace had picked me up, and when she dropped me off at home I invited her in. We sat on my couch, sipped tea, talked, and reflected on the dance. "I really wanted to share the last waltz with you," I said. "I'm sorry that didn't happen."

"If you hum the music," she said, "you can still have the last dance."

We stood up. I took her hand and slowly hummed "Iowa Waltz" as we moved softly around the room. I pulled her closer and felt my heart melt in my chest, overcome with a sensation I'd never experienced with any other woman.

An inward clarity whispered to my heart that she was the one. Yes, she was absolutely the one! She was the woman I was supposed to be with, with whom I was destined to spend the rest of my life. In the sweet, small circles of a waltz hummed softly in the ear of his dance partner, one Victorian Man had found his Dulcinea.

In my tent along an Arizona highway, two days' walk from Phoenix, I imagine that when I again fall asleep my dreams will guide me back to that moment with Grace, guide me to a realm of time and space where our waltz never ends, where her head forever rests against my shoulder, her hand fitting perfectly in mine, my other hand around her waist, holding her close, dancing slowly, spinning ever onward beyond turmoil, beyond pain and suffering, beyond eternity itself.

Despite the physical fatigue of marching and the emotional battering from other marchers, despite the traffic noise, despite missing Grace, I'm at peace in this moment. Shira is asleep, and I'm content knowing she will be there after I fall asleep and when I wake up.

In addition to sharing a tent, Shira and I usually march together, almost always with Steve who, like us, prefers to set out before other marchers. We've become an efficient, close-knit marching trio, united by our quick pace, our eagerness to set out while the day is still fresh and young, and our discontent with the level of drama afflicting the March community.

We wake up well before others, usually between 4:30 and 5:30. We are fast and focused as we pull up our tents, stow our gear, grab breakfast, and pack a lunch. We're gone within 30 or 45 minutes, before most marchers are even awake.

Some are bothered by our early departure. They want us to wait until all who intend to march (increasingly only about half the group) are ready. Steve, Shira, and I argue that a few disorganized marchers who take two hours to get ready should not delay everyone else. I suggest that, with the exception of those whose job it is to stay behind and break camp, these marchers need to be better prepared and more responsible. We should all head out by 6:00, perhaps earlier when the day will be hot or the mileage long. Furthermore, I remind others that I need the early start to accomplish the day's fundraising and other work in my role as director.

In quiet moments, when I'm able to analyze my actions and motives more fully, I understand that my desire for an early departure is also about avoiding those marchers who have become hostile. It is both sad and remarkable how quickly and completely the feeling of good will within One Earth Village has collapsed. I still get along with most marchers, but the earlier camaraderie of the community as a whole has been shattered. A few disgruntled marchers have created an environment that makes me want to be elsewhere, anywhere, although mostly it makes me want to be home where even my political opponents treat me better.

So I work and I walk. As much as possible, I work while I walk. I've become competent at making calls, answering emails, and sending text messages without missing a stride or spraining an ankle.

But sometimes, it's impossible to work while I walk. Rain, wind, traffic noise, dead zones — I can't operate my phone in such conditions. Sometimes I'm relieved, happy for a reason not to work. I talk and joke with Steve and Shira. We pause to snap a quick photo of something beautiful or unexpected. We stop for short breaks or to talk about climate change with people curious as to why we walk.

Shira loves theatrics and finds silly reasons to take photos and videos. Once, I film her pretending to gnaw on elk bones; another time, she juggles elk turds. On another occasion, she poses for a photo pretending to dangle off a cliff.

There is usually little or no shade as we walk along the highway across Arizona. When scouting a spot for a break, shade is the highest priority. Sometimes, our only relief from the sun is a telephone pole, and the three of us sit, lie, or stand in a row in its thin but welcome shadow.

In the desert and mountains, without cafes where I can stop and work, I now try to get through the marching day as quickly as possible so I can work from camp before most marchers arrive. Where I once was the last marcher to roll into camp, I'm now often the first.

We arrive in Phoenix on Steve's 60th birthday. A few of us celebrate with him at a restaurant. We also celebrate the completion of 400 miles of marching.

Our rally in Phoenix is small — about 100 people clustered in a tight circle under shade trees in a downtown park. The highlight of the rally is a blessing ceremony led by a Navajo elder who entrusts two small turquoise stones to the March. He hands one to Miriam and one to Steve, telling us that the stones embody the concerns of the Navajo community for Earth and water. He asks Miriam and Steve to give the stones to our country's leaders when we meet with them at the end of the March.

Overall, there's been less receptivity to our message in Arizona than in California, even though Earth's changing climate will hit Phoenix harder than most places. It's great to see Arizona's progress on solar energy, where there's now enough installed solar capacity to power 339,000 homes. That ranks Arizona second in the nation.[1]

Yet in the New Climate Era, renewable energy cannot save Arizona, particularly Phoenix. Increased competition for a declining supply of water suggests a nonviable future for Phoenix. Like a physician trying to assure a dying patient that her condition isn't entirely dire, the *Third U.S. National Climate Assessment for the Southwest* almost certainly understates the world of hurt coming to Phoenix: "Snowpack and streamflow amounts are projected to

---

[1] http://www.seia.org/state-solar-policy/arizona

decline in parts of the Southwest, decreasing surface water supply reliability for cities, agriculture, and ecosystems."[2]

We have home stays in Phoenix — our first since Claremont. I stay with Chuck and Linda Necker and talk with Chuck about the future of Phoenix. He concurs that the New Climate Era will likely punish Phoenix badly. While no place in the world is immune from climate change, I tell Chuck that he and Linda might want to purchase a house in rural Iowa, where the real estate remains cheap and the water supply abundant. "But buy before the vast migration of Americans from the Southwest, Florida, and low-lying coastal regions begins," I tell him. "My prediction is Iowa's population of three million will reach 30 million by 2100, if not sooner."

The March strikes out from Phoenix on the morning of April 8 for a 13-mile day that takes us beyond the city's suburban edge. Steve, Shira, and I detour along quieter roads so I can better manage a staff call while walking. Our route takes us through a park with sandstone-sculpted rocks and hills. I walk cautiously, as I'm certain a rattler lies in wait under every large rock we pass.

I walk hard and fast today and arrive at camp by 10:00 a.m. Marching is the easy part of today's workload as I've agreed to help Sarah scout the next 75 miles of the route. Three campsites between Phoenix and Payson have yet to be determined.

As Sarah and I drive north and east toward Payson, I joke that she's achieved the distinction of the second-most maligned person on the March. "The harder we work, the more flack we get," I muse. "Do you think marchers would be happier if we did a shitty job at our assigned duties and just hung out at camp more often?"

"Oh, I don't think we can win," mulled Sarah. "Best just to keep plugging away and gittin' 'er done."

The day is long but productive. It's an odd feeling, to drive the route I'm about to walk. We score solid leads on three campsites and get a sense of what to expect in Payson. At 10:00 p.m., we arrive back at our campsite to discover there'd been a mix-up. The camp I walked

[2] Third U.S. National Climate Assessment for the Southwest

to this morning was set up in the wrong location. One Earth Village has since moved a mile farther down the road. So as not to miss a step, I ask Sarah to drop me at the original campsite. Tired but satisfied, I walk the final mile to the new location under a night sky filled with bright stars and howling dogs.

Tomorrow, we set out toward Payson, elevation 5,000 feet — and then over the Mogollon Rim to 8,000 feet, where we'll again brace for cold nights and potentially wild weather. Still, I'm reasonably sure we'll never experience a day with conditions as rough as our first day's march in Los Angeles.

The high desert plateau beyond the Mogollon Rim would soon prove me wrong.

CHAPTER 12

PAYSON

"In the empire of desert, water is the king and shadow is the queen."

— MEHMET MURAT ILDAN

It's a six-day walk from Phoenix to Payson, a gradual climb that becomes more beautiful the higher we go. Saguaro cacti are abundant, their tall trunks and outstretched arms a source of amazement and amusement to those of us unfamiliar with the desert.

I pose for a photo, pretending to kiss the bristly trunk of a towering 18-footer. Steve smiles and says in his distinctive Southern accent, "They look like they just wanna give you a big hug." Jane Kendall, a marcher from New York City, offers a different perspective. "With their arms up in the air like that, they look like they've just been robbed."

The range of the Saguaro cactus ends at 4,000 feet. In one day's walk, these stately, spiny huggers — or mugging victims, depending on where you're from — give way to juniper trees, which although less distinctive are a better source of shade for a hot marcher.

As we push through cooler temperatures upwards toward 5,000 feet, pine trees mingle with the junipers, turning the mountainsides green. As a prairie-dwelling lover of all things verdant, I feel more at home here than at any point in the March.

As we approach Payson, a conflict ensues that threatens to undermine the commitment of the eight of us who have walked every step of the way. Ray Spatti and the local Payson welcoming committee have worked tirelessly to bring the City's leadership on board to support our visit. But Highway 87 is the only road into Payson from the south. With no shoulder and many sharp turns, it is dangerous if not deadly. Ray pleads with us not to walk the final ten miles. "Locals have offered rides to all the marchers," says Ray. "Be safe. Take a ride. It's only ten miles."

City officials weigh in. They practically order us not to walk to Payson. One frustrated but well-intentioned resident says, "If you insist on walking, we'll contact the funeral home and tell them to expect new business."

I stand firm. "We're committed to marching. We've come nearly 500 miles and have found a way to walk on or around every stretch of dangerous highway. I don't yet know how we're going to get to Payson, but I promise we'll get there by walking and without getting killed. Taking a ride is simply not an option."

Trying to lighten the angst of our hosts, I add, "So tell the funeral home we're sorry to disappoint them, but there'll be no new business for them tomorrow, at least not from the March."

Mack has taken a leadership role in emergency re-routing and this situation presents one of the most challenging yet. He and I carefully study the map on either side of Highway 87. We discover a back road that leads west out of Rye, the town just before Payson. The road dead-ends but leads to a series of trails that appear to carve a path north through the mountains.

The route is difficult to piece together but it works perfectly. We exit the wilderness just beyond the dangerous stretch of highway and accomplish something that happens infrequently on the March: everyone is happy.

Our visit to Payson is a towering success. The Boy Scouts lead the procession from the city's edge to our campsite on the soccer field at the Payson Christian School. That evening, Mayor Kenny Evans and other city leaders join us for dinner and a conversation on climate. *The Payson Roundup* writes a favorable story. The experience is reminiscent of the enthusiasm we relished at our kick-off in Los Angeles six weeks earlier.

"This is why we walk," I tell Ray. "People are interested in why we're here and are willing to talk about climate change." I remind Ray that one person can indeed make a difference, and that he is proof of this truth.

We have a day off in Payson. That evening, after working for a good chunk of the day at Dimi Espresso Coffeehouse, I meander through a residential neighborhood on my way back to camp. I'm impressed that there are no grass lawns but instead some combination of sand, rock, cacti, and other desert plants. As I peer through the dusk, I see what appears to be a green lawn in the distance. "Climate slacker," I think, as I wonder how much water is needed to maintain this patch of green grass in the desert.

A little closer, and I notice large shapes moving on the lawn. Javelinas! A herd of five or six wild pigs are rooting up the only green lawn in the neighborhood. I pass them quietly, not wanting to disturb their important work as they push back against climate change in their own small, piggish way.

We leave Payson on April 15, walking through a very different landscape than most of what we've experienced in Arizona. For me and other marchers who miss the familiar comfort of green, the towering Ponderosa pines and other conifers are a welcome respite from the sparse sand-scape of the Sonoran Desert. We enjoy two beautiful, sheltered campsites, though the nights have again grown noticeably colder.

On April 17, we camp just south of the Mogollon Rim. For over a week, we've watched the Rim draw closer with every turn of the road. I'm reminded of other mountains I've climbed, how the excitement and expectation build as one approaches the ascent to the summit. From a distance and now up close, the Rim looks and feels like any other mountain — sensational, yet familiar. I'm puzzled as to why it's called "the Rim."

The road today is steeper than any we've encountered. High altitude forces us to stop often and catch our breath. I admire the spectacular vista to the south — each peak shorter than the one to its north, the mountains diminishing in grandeur as they roll toward the parched desert below, like visual claps of thunder fading beyond the horizon.

As always, we walk toward oncoming traffic, which today is light. Still, the road is dangerous with frequent sharp, blind curves. Drivers aren't expecting walkers and there's little we can do to warn them of our presence. We hear vehicles approaching well in advance, giving us time to hug the cliffs that line the road.

Reaching the Rim is exhilarating. It's as if we've just scaled Mount Everest. I expect to see more mountains north and east. Instead, the land suddenly and dramatically flattens. "Well, I guess those Bible thumpers were right," I joke. "The Earth is flat, and we just crawled over its edge."

It does indeed feel as if we've scrambled up the side of the Earth and now triumphantly survey a planet lush with pine trees. I suppress the knowledge that, in a few miles, the pines will yield to scraggly junipers, then to high desert, where it can be cold, windy, and often inhospitable in spring.

We rest just beyond the Rim at 8,000 feet. I'm utterly exhausted. Shira and Steve found the ascent challenging, too, but it hits me hardest.

"This altitude is kicking my ass," I say. "All three of us hail from flat country, but you guys seem to be handling it better."

"Maybe it's not the altitude," says Steve. "Maybe it's all that weight you've dropped."

"Well, I've been feeling sluggish since Payson, and have dropped 24 pounds in just six weeks," I reply. "You've lost a bunch of weight too, right?"

"Yeah, twenty pounds. But that was weight I could afford to lose. You, not so much," Steve laughs.

I consider Steve's viewpoint. "Nah, you're no doctor. It's the altitude," I conclude defiantly.

But I'm uneasy. I've always listened to my body. And right now it's telling me something's wrong, something I need to figure out soon or pay the consequences.

A MEMOIR FROM THE GREAT MARCH FOR CLIMATE ACTION

CHAPTER 13
FOOD

"The belly is an ungrateful wretch, it never remembers past favors, it always wants more tomorrow."

— **ALEKSANDR SOLZHENITSYN,**
ONE DAY IN THE LIFE OF IVAN DENISOVICH

"Maybe a higher power is telling us we should've taken Easter weekend off," I suggest to Steve and Shira as a cold wind nips at my ears and fingers. Tomorrow is Easter Sunday and we'll arrive in Snowflake — a small town with fifteen churches, more than half of them Mormon.

Walking through a deeply religious community on Easter Sunday is not our finest organizational accomplishment. Erica and I had spent hours talking with church leaders and City officials about places to camp, but met with solid rejection.

"Yeah," says Shira. "We should have stayed a second night at Camp Shadow Pines."

Oh, would that have been nice! At Camp Shadow Pines, we slept in beds. One well-wisher brought us cookies, another hot chocolate. We peed in ceramic toilets. It was heaven.

But we have to keep moving. Our New Mexico coordinator, Ahni Rocheleau, has been hard at work with rallies scheduled in Albuquerque, Santa Fe, and Taos. So we rush along at three miles an hour.

Today opens with a six-mile walk from Camp Shadow Pines to Heber. I skip breakfast, knowing John Abbe is cooking and that he'll probably make oatmeal — or as I've come to call it, oat glop.

John's concoction is so thick it could be used as mortar. The pot is large and he forgets to stir it. The bottom burns badly. John doesn't want to waste anything so he stirs the thick layer of burnt oatmeal in with the rest, imbuing the entire pot with a rich, carbon flavor. This culinary atrocity is further enhanced with fistfuls of raisins, cooked until they swell to three times their size, resembling engorged ticks.

Choking down this food substitute was less oppressive when I would day-dream about a second breakfast at an imaginary café ten miles down the road. While fantasizing about a plate of eggs, bacon and pancakes, the tick-like raisins coated with burnt oatmeal would slide down just a wee bit easier.

Today, the temperature hovers just above freezing as we escape from Camp Oat Glop in the nick of time. A couple hundred yards into the walk I realize I've forgotten my gloves.

"I'll be fine," I say. "No need to go back," as I wrap a coat sleeve around the hand holding my walking stick. I tuck the other hand in my pocket and switch hands every minute or so. Both grow numb with cold.

Two hours of discomfort are rewarded with the most satisfying breakfast yet. We stop at June's Cafe and I order the special: two eggs, two pancakes, bacon, sausage, and hash browns. Still hungry, I order two more pancakes, then two more on top of that.

Our waitress is amazed. "How does a skinny guy like you put away that much food?" she asks.

"I have a walking problem," I confess. "Seven weeks and 600 miles ago, this skinny guy was noticeably heavier."

Despite the huge breakfast, my fatigue persists. The remainder of the day's 15-mile march is difficult.

I call my friend, Dr. Charles Goldman, a physician and frequent guest on my talk show. "What do you think about this dramatic weight loss, Charles? Is that why I feel so lethargic? Or is it just cause I'm a flatlander adjusting to altitude?"

"No, from what you describe, it's not altitude. Your problem is you're dying," says Charles, who took a personality test in med school and ranked near the bottom in empathy.

"Thanks," I reply. "Been nice knowing you."

"Ok, you're dying, but you've got an exit strategy. Your body's consuming itself. You've burned through the fat, which you really didn't have. You've burned through the carbs, and now to keep pace with your caloric output, your body's eating its proteins. You're basically consuming your muscles, which is why you feel like crap. Keep it up and, yeah, probably before you get to DC, you'll die," Charles concludes with a chuckle.

"I see you've been working on your bedside manner," I shoot back. "But seriously, what do I do? You gonna prescribe some overpriced medicine?"

"Yes. Your prescription is to eat prodigious amounts of food, especially meat," advises Charles, who himself is a practicing vegan. "And yeah, by the time you get to DC and tally up the cost of all those diner meals, pharmaceuticals will seem like a bargain. But at least you'll still be alive. Maybe."

Charles's diagnosis explains my recent bizarre craving for flesh, too. Occasionally, we'd walk past a herd of cattle, and where I'd normally think, "Nice cow," I'd find myself muttering, "Nice steak." At other times, I'd see a squirrel running up a pine, and ponder how many tree rats it would take to make a tasty pot of stew.

"Next, I'll be drooling at the sight of roadkill," I'd muse.

The March's vegan crusaders have pushed hard against the evils of flesh. Meals on the March featuring meat have grown scant. I try to make do but it isn't working. The vegan cooks serve a lot of couscous. Once, I was so famished I had a third heaping bowl of the stuff. Despite my hunger, I couldn't choke it down. It felt like I was eating granulated cardboard.

Similarly, though I love a nice salad, I now find little culinary or caloric joy in lettuce and kale. Granola and yogurt, my usual breakfast staples back home, are profoundly unappealing. Normally, I love peanut butter. Now, I can't stand it, though perhaps eating nearly 50 peanut butter sandwiches during the first four weeks of the March explains why.

Taking Charles's advice to heart, after wolfing down the equivalent of two large breakfasts at June's Cafe, I vow every day to stop at a greasy-spoon pharmacy, if I can find one, and diligently take my medicine of eggs, bacon, sausage, and pancakes.

Two days later, having survived our walk through the holy city of Snowflake, smitten neither by God nor man, we meet Tim and Mary Windwalker. They donate strawberries, hot dishes, and guitar strings to the March. Mary asks me what else they can do to help.

"Got an Internet connection and phone service I can borrow?" I ask.

"Sure, hop in the car and come over to our place."

Their home is like none I've ever seen, an "Earthship" that Mary and Tim built from adobe and old tires in 1994. Solar panels provide most of the power, although the design itself takes care of much of the heating and cooling needs.

Everything about the interior is comfortable. Art is everywhere. A dragon carved from a hibiscus plant sits next to a poinsettia. Veins of stained glass run across the walls. A lotus flower painted with soy-based stain decorates the entry-way floor. A jungle-like greenhouse occupies the southern wall of the Earthship.

Tim and Mary make shoes for a living. Their shop runs entirely on solar. "We've made 4,000 pairs of shoes with equipment powered entirely by the sun," notes Tim. "Fossil-fuel-free footwear," he smiles.

Like me, Sarah's work demands Internet and phone service. She shows up a bit later and we spend a few hours catching up on March work. Mary invites us to stay for dinner: barbecued, wild-caught salmon marinated with raspberry chipotle sauce. There's a salad of beets, carrots, and pinion nuts. Organic red wine provides a rare treat.

The night is cool but not cold. We eat, talk, and relish this moment of life under a full desert moon that drifts slowly over the White Mountains to the south.

"Look at that," says Tim. "An unobstructed view, free of power lines, cell phone towers, wind turbines, or bright lights. There aren't too many places you get to see that."

True enough. It feels as if Sarah and I have stumbled on a slice of paradise. Yet all is not well in paradise. Tim talks about how every year the desert creeps a little closer.

"When the desert reaches your doorstep, what will you do?" I ask. "How will you adapt?"

"We'll leave," said Tim. "What else can we do? Without water, what we have here can't exist."

Tim speaks matter-of-factly about what climate change and impending desertification mean for him and Mary. Maybe they've already come to terms with it, but I feel a deep sense of sadness that all this effort and beauty might be abandoned.

Mary invites us to stay the night. Sarah needs a break from camp and accepts. "As much as I'd like to stay, I'd better get back," I say, knowing that the grumbling against me has grown worse since Phoenix. It's not just about money and staff. I can't put a finger on it, but something else is brewing, and I sense there's new turmoil and contention to come.

On April 22, Earth Day, we arrive at the home of Mark Boyko and Karen Abbott near Concho. They graciously offer us a two-day break from wilderness camping. Like many people in eastern Arizona, they live off the grid, their home powered entirely by solar and wind energy. We find ample space to pitch our tents, yet due to the howling wind and cold temperatures, Mark and Karen let marchers sleep in their spacious Quonset hut. They also invite us to use their shower.

Lala prepares the best meal of the March: a multi-course, Mediterranean-style dinner featuring ratatouille, baba ganoush, tabouleh, falafel, and hummus. The only thing missing is meat. I play my accordion while Lala and others cook. It's a peaceful, happy occasion. Some marchers dance. A few sing along. Lala requests "Never on a Sunday," one of her favorite tunes. I haven't played it since I was a kid, yet it comes back to me and Lala is thrilled.

Lala insists that tonight's meal is not just about food but about presentation. The meal is served at tall, round tables. Marchers and our guests stand at the tables and move from one to the next, socializing with new dining companions throughout the meal. This constant switching of partners strikes me as the culinary equivalent of a contra dance.

After dinner, I'm confronted with the reality that the March's financial situation is worse than I thought. Earlier in the day, Shari informed me that we needed to raise $5,000 by the end of the week in order to meet our commitments. I know I will have to work hard on our day off and avoid distractions. Kim and some of the other marchers won't be happy that I'm not socializing, but I see no way around it.

Mark and Karen have a spare bedroom. They let me hole up there during our two-night stay. This works well. I'm able to raise $5,000 and meet our financial shortfall. But it comes at a price, as I'll soon discover.

"Yeah, staying in that nice, comfy bedroom. That was probably a bad call," says Shira as we march together the next day. She spent the night in the Quonset hut with most of the other marchers. "They say you're 'profiting' off the March. They say you're mismanaging our funds. Now you're a pampered elitist," Shira says with a laugh.

"Well," I assure her. "I try to do something stupid every day. It keeps me young."

"Besides, look at this!" I mutter, waving my phone in the air. "Dead. There's no Internet or phone service out here at all today. Instead of being productive while I walk, I have to listen to you and Steve babble," I laugh.

But I know Shira's right. Although the occasional opportunities to cloister myself and get work done are necessary, especially when finances are tight, I'm fully aware they provide my detractors more raw meat to gnaw on.

I put that out of my mind. Spring in the high desert has turned harsh, threatening to transform the March into an episode of *Survivor*. Nights are much colder, with temperatures dipping to 20°. Days are cold, too. We march bundled in layers, counting on the exertion of walking to keep warm.

On such days, camp life is miserable. Cooking, eating, and using the Eco Commodes are uncomfortable. I'm concerned that marchers who are less cautious and less experienced with the outdoors may be at risk from frostbite or hypothermia.

As it turned out, I should have been more concerned about myself.

CHAPTER 14

HYPOTHERMIA

"Nothing burns like the cold. But only for a while. Then it gets inside you and starts to fill you up, and after a while you don't have the strength to fight it."

— GEORGE R.R. MARTIN, *A GAME OF THRONES*

Nature and I have always been lovers. As a kid, after my bus ride home from the torture chamber of Catholic grade school, Nature would soothe my wounds. Her love helped me forget, at least for a time, the physical and verbal beatings by bitter, barren nuns too old to teach and too senile to care.

Nature helped me forget the time I was slapped hard in the face for the crime of waving to a friend through the window of a classroom door.

She helped me forget the time I was punched in the stomach on the playground by another student, completely unprovoked, and then brought to the principal's office to be spanked.

She helped me forget that my best friend in third grade was beaten nearly every day — and that Sister Antionette, the nun who beat him, held him back a grade so she could beat him again for another year.

While Nature couldn't stop the abuse, she provided a balm where none was offered.

I'd jump off the bus, dart home, tear off the tie that hung like a noose around my neck, and run into the forested arms of my lover. With Nature, I always felt unconditional acceptance.

We would flirt with abandon. I teased her, dared her to love me as much as I loved her, dared her to love me to the very edge of self-destruction. We played the games lovers play. She always won. "But only because I let you win," I taunted.

When the moon waxed full and flooded the salt marsh near my childhood home, I'd venture out before the tide's crest, eager for a playful fight. I'd find my way to some thin sliver of ground, a spot slightly higher than the land around it. I'd lay claim, drive a stake into the mud and attach a rag to its top to let Nature know what king now reigned here. Then I'd await my lover's watery wrath and the inevitable advance of the sea.

The tide would push in gradually, conquering a few inches of land at a time, until my kingdom was reduced to a tiny island. At the last minute, I'd heroically leap to dry ground, abandoning my claim, managing a dramatic escape from a cold, briny death.

As I ran home, I'd glance back at the wreckage of my kingdom and watch Nature submerge all but my rag in her ocean's cold, healing waters. "See," I'd shout. "I let you win again."

Our games were particularly fun in winter, when ice flows provided even greater opportunity for juvenile foolishness. I never came close to drowning, but twice filled my boots with icy water and returned home shivering, cold, and feeling very much loved.

At Marlboro College in Vermont's Green Mountains, my affection for Nature remained constant — and my devotion to playing stupid with bad weather escalated. Once, on the coldest night of the year, when the mercury dropped to -30°, four of us ventured out for a five-hour cross-country ski excursion as a watchful full moon illuminated our snow-covered path.

Sub-zero weather is no place for gloves. It's mitten weather. Yet one skier, Kate, insisted on gloves. Thirty minutes out, Kate complained of pain in her fingers. Fortunately, we'd brought an extra pair of mittens. But when Kate stopped to put them on, her white fingertips signaled the onset of frostbite. I quickly stuck her hands on my bare chest under five layers of clothing, and held them there until they recovered. The mittens did the trick. Kate managed the adventure without further incident, and I had my first girlfriend.

Well, second after Nature.

A MEMOIR FROM THE GREAT MARCH FOR CLIMATE ACTION

Internet service between St. Johns, Arizona, and Zuni Pueblo, New Mexico, borders on nonexistent. This morning, a man offers to let me use his wifi and I dare not refuse. I work at his home while other marchers walk. My plan is to complete the 12-mile trek this afternoon.

At just under 20°, last night matched our coldest temperature yet. By noon, the mercury has warmed into the upper 30s, though this benefit is mitigated by a wind that blows stronger as the day progresses.

I set out shortly after noon, just as other marchers are finishing, in a veritable gale with a tailwind raging at 40 miles an hour. The wind bats me around like a cat toying with a mouse. Gusts reach 55 miles an hour. When they slam me from behind, it's impossible not to run. I laugh as I'm pushed along helplessly, as if in some kind of circus funhouse.

Sometimes, the wind veers slightly right and tries to push me into the ditch. It quickly shifts left, sending me toward the highway. Then it returns to a straight tailwind, and I run, laugh, surrender, give thanks for my comparative good fortune.

"Ok, so this ain't so bad," I reason. "With a tailwind pushing me along, today's 12-mile stretch will zip by in no time — if I can avoid being blown into the ditch or an oncoming car."

But after two miles, the wind shifts dramatically. Now it roars out of the west — a crosswind, and colder. I lean hard to my left to keep from being pushed into the road. Ten miles of this will be tough.

A mile later, the fierce wind brings hail, then rain. At 40 miles an hour, the raindrops sting like a barrage of icy needles. I try to protect my face with my gloved hand, with some success.

By the end of another mile my jeans are wet. Legs warmed by walking become cold quickly as the rain soaks through. The glove guarding my face is wet, too, and the fingers of my left hand grow numb.

At six miles, the rain changes to snow — a wet, ill-tempered snow that pummels me mercilessly. Before long, everything is wet, even the shirt under my raincoat.

I think back to my time on the marsh as a kid. I recall Nature's harsh side. Despite the harshness, I felt loved. "Nature, ole gal, this doesn't feel like love," I shout into the wind. "Are you pissed at me?

Don't you love me anymore? Is this the day you finally break my heart along with my body?"

The wind and snow are relentless. There's nothing I can do but keep my legs moving and my mind focused. The March support vehicles are at camp and my phone won't work in the rain. I'm on my own with six miles to go in the most inhospitable conditions I've experienced yet. "Maybe if I walk fast and hard, I'll generate enough heat to offset the cold," I strategize.

Five miles to go, and I only feel worse. My lips are cold. Walking fast becomes difficult, and though I want to push hard to regain warmth and get through the walk as fast as possible, my body doesn't respond, doesn't want to move.

"Mind over matter," I tell myself. "Mind over matter. Think happy thoughts." Grace enters my head. Every day, I look at the photos she sends. They are both artistic and provocative, wholesome yet arousing. One photo shows the bottom half of her legs as she sits in the bathtub. In another, she's lying on pavement, looking serious, her hand buried in her rich, black hair. In another, she stands in front of a mirror, trying on a pair of reading glasses, sporting a pink tank top, hand on her hip.

The image of Grace becomes a mantra. I focus on her like a meditator hones in on a candle flame. I think about Grace in the present, wonder where she is, wonder what she's doing. Mantra, or daydream, it's all good, all a distraction from the discomfort of the cold. I'm transported to a place in my mind where life is beautiful, and I feel more able to manage such brutal conditions.

A particularly sharp gust of wind snaps me out of this mental respite, thrusting reality back into my face. I'm again aware of the extent of my deteriorating condition. I desperately search for some place to hide from the wind and snow.

With four miles to go I come to an overhanging rock — a godsend that provides shelter, albeit little comfort. For the first time in two hours, I'm not under attack by wind, hail, rain, and snow. I catch my breath, try to get comfortable, and hope that this break will allow me to recover and finish the day. I take from my pocket the locket Grace made for me just before I left Des Moines, the locket I've carried with me every mile of the March.

I smile as I admire it and realize it's been several weeks since I've looked at it. Under the locket's glass cover is a map of Iowa. Inside is a wheel of sixteen different colors of felt, each like a sliver of pie, all converging in a center of blue and yellow.

I study the locket more closely than at any time during the March. Each sliver of felt symbolizes something Grace and I shared, something she loved about the times we spent together. Music. Gardening. Food. A pinkish piece of felt stands for the baked grapefruit I once made for breakfast — I remember how Grace was visibly moved the first time I served it to her, with each segment cut and ready to eat.

Other strips represent sunsets, mangos, okra, and red wine that I would serve in tea cups from my family's farmhouse in Ireland. There's a strip of blue felt, symbolizing the little urban "river" that runs through two blocks of downtown Des Moines, where Grace and I enjoyed our first conversations.

Crouching under this rock, where the jaws of the cold desert wind snap but can't bite, I feel transported to a better place. But inevitably, I notice that I'm no warmer now than when I arrived, perhaps colder even. I can't tell.

I put away the locket. I can't linger. I have to keep moving. I leave my dry shelter, a source of comfort not so much because of any warmth or physical relief but because of the opportunity to lose myself in Grace's locket. I move slowly. I feel the cold numbing not just my arms and legs but my chest as well. I keep moving, one step at a time closer to camp, almost in a trance.

Though the wind continues to rage, the snow has stopped. Eventually, I see the camp not far ahead, less than a mile away. A car approaches. It pulls over. It's Bob Gardiner, Izzy Mogelgaard, and John Jorgensen. "It's brutal out here today, even worse than last night," says Bob. "Our tents are taking a beating, so the three of us are gonna get a cheap hotel in St. Johns. You wanna join us? You look miserable."

"Well, if you can hang on while I finish the last half mile, I'll join you," I say through chattering teeth.

"Forget about the last half mile and get in the car," insists Bob, who then recalls my stubbornness. I start walking toward camp. "Never mind," he says, "We'll just meet you at the camp entrance in 10 minutes. And we'll grab your gear from your tent."

"Thank you," I say. "And ask Shira to join us, too. There's always room for one more climate marcher on a hotel floor, right?" I joke. "I don't want her freezing to death because her tent buddy abandoned her. Seriously. It's gonna be a rough night."

Shira opts to gut it out in the tent, how I don't know. Fifteen minutes later, I'm shivering almost uncontrollably as I slide into Bob's car. He cranks up the heat, which makes the other passengers uncomfortable and does nothing to relieve the cold that seems embedded in my very bones. At the hotel, it's a full two hours before I'm warm again. I realize that if I'd stayed at camp, my condition probably would have required medical attention. Instead of a hotel, a hospital emergency room would have been my overnight lodging. I escaped the high desert's wintery wrath just in time.

But not all marchers are as understanding as Bob, John, and Izzy. Michael Zambrano, a close friend of Kelsey's, sends a message: "Hey Ed, I hope you enjoy your night in a hotel room while the rest of us suffer." I want to scream, want to strangle him, want to yell, "Have you no clue what kind of weather I walked in? Does it not matter that I spent the morning raising money so you could march? Would you rather I'd stayed at camp until the ambulance showed up to haul me away with hypothermia?"

Michael has become Kelsey's lieutenant in her campaign to oust me as paid director. He's been more and more unfriendly toward me since Payson. His words cut almost as harshly as the day's wind and rain.

I make my bed on the floor of the hotel, grateful to be warm and dry, feeling fortunate to be alive, thankful for Bob's kindness. But sleep doesn't come easily. I'm nervous about marching tomorrow, given the forecast for continued harsh weather. Furthermore, Michael's cruel remark gnaws at my gut.

A MEMOIR FROM THE GREAT MARCH FOR CLIMATE ACTION

CHAPTER 15

ZUNI ENCHANTMENT

"To encounter the sacred is to be alive at the deepest center of human existence… At Devil's Tower or Canyon de Chelly or the Cahokia Mounds, you touch the pulse of the living planet; you feel its breath upon you. You become one with a spirit that pervades geologic time and space."

— N. SCOTT MOMADAY

The wind continues to blow hard during our final day across Arizona. The air is at least dry but remains cold, windy, and without comfort. Instead of rain and snow it blows sand — a sand like none I've ever seen, fine as dust and almost imperceptible until you feel it clogging your nose, irritating your eyes, scratching your throat.

Worse yet, Sarah is unable to beg, schmooze, or conjure up a campsite. Seeing no option, she commandeers a thin strip of scrubland sandwiched between the highway and a fencerow. With neither a DOT permit nor the landowner's permission, I fret about the possibility of a midnight eviction.

Our camp is normally compact. Tonight it stretches over 100 yards along the highway. The wind rages so hard it takes a team of marchers to set up a tent. Two hold the door open while another dives inside, lying spread-eagle to keep the tent from blowing away. We then secure the corners and sides as best we can in the hard-packed surface, attaching the fly to the ground as tightly as possible.

Yet even our tents provide no escape from the high desert's violent thrashing. Mysteriously, the sand finds its way through the tent walls. By morning, a layer of brown-yellow dust covers everything, including our heads, faces, and hands. I mumble to Shira that the floor of my cheap hotel in St. Johns the night before was a slumbering paradise compared with this dusty niche of hell.

The next day, we enter New Mexico where the Zuni people have agreed to let us camp on the grounds of the Zuni Correctional Facility for Juveniles. But with cold, windy weather persisting, officials insist on putting us inside the center instead.

For tent-dwellers accustomed to sleeping quarters the size of a compact car, the correctional facility's twelve jail cells are prime real estate. All are claimed by the time I show up, and I settle for floor space in the common area with other late arrivals.

"This is the first time we've filled the place without arresting a single person," jokes one of the Zuni officers.

Beyond the luxury of spending a night in jail, staff provide hospitality above and beyond anything we expect. This evening, they launder our clothes and serve a delicious dinner of standard American fare mercifully devoid of couscous. After dinner, we pile into prison vans. Guards drive us to the heart of Zuni Pueblo to watch traditional dancing in the plaza of the old Catholic church. We're told that, in no uncertain terms, we're not allowed to take photos or shoot video. The documentary crew is particularly disappointed, but everyone complies.

Arriving at the village, I'm unprepared for the uniqueness of the experience. I feel as if I've been transported to a remote country in a simpler time. Zuni Pueblo is home to 11,000 people. It's cozy and walkable, the houses small and simple. Adobe ovens sit outside most homes and many are in use as we pass. The director of the Zuni Pueblo MainStreet program, Wells Mahkee, explains that people are baking bread and making stews and roasts that will cook through the night.

Towering above the Pueblo is the flat-topped mountain called *Dowa Yalanne*, a high mesa sacred to the Zuni that outsiders are not allowed to visit. I admire it from the window of the prison van as the evening sun breaks through the clouds, illuminating the mesa's steep escarpment. It's a stunning sight, the mountain's spiritual and physical power manifested in the setting sun's brilliance. I give thanks that such places exist to nurture mind and spirit.

As we walk from the van to the church, the drumming and singing grow louder. What strikes me most is that nearly everyone is speaking Zuni! Here, on a reservation also unique for its lack of a casino, the language of daily life and commerce is the Native tongue.

At the church, we're quietly ushered up a winding staircase. The air is tangible with excitement as we reach the rooftop, the plaza below packed with dancers. The moving display of color is like looking through a kaleidoscope, the bright patterns changing as dancers move around the plaza. The drumming and singing are riveting, hypnotic. The experience is a sensory overload, dizzying even, as if we've entered an altered state of consciousness without the assistance of the usual plants or drugs.

One mask stands out — drab, Earth-colored, disconcerting, a bit frightening even. Six or seven shirtless middle-aged and older men wear this mask. The mouth is a round hole protruding outward from the face, with unnaturally large and hollow eyes. The mask feels not quite human, not entirely right, as if its bearer had crawled out of a deep, forgotten underground cavern to taunt and unnerve other dancers.

"Mudheads," Wells explains to me later. "That's what we call them. They're a physical representation of what happens when you procreate with someone from your own clan. They're a symbolic warning against incest. They're scary, but they're also clowns. During intermission, Mudheads entertain the crowd, do funny antics, tell stories."

We aren't on the roof long before the prison guard says we have to leave. I regret not seeing the Mudheads' halftime performance. But when your driver is a prison guard and says it's time to return to jail, you obey.

Back at the correctional facility, I ask the guard what time the lights go off. "Are you kidding?" he laughs. "This is a jailhouse."

"But we're good girls and boys, didn't do nuthin' to nobody,"

I plead with a grin, hoping he'll make an exception. The guard just laughs again and I feel a bit silly for asking.

Marchers lucky enough to have cells use cardboard, sheets, and tape to cover light fixtures. In the common area, we don't have that luxury. Worse still, the jail's lively acoustics amplify every sound. Nearby sits the "bathroom" — no door, just a half wall in front of a toilet. Every time a marcher gets up to pee — and this happens often — the resounding flush jars me awake.

Another sleep-disrupting sound is amplified by the jail's acoustics, as two marchers have sex in their cell. I laugh quietly, imagining this also has to be another first for the correctional facility.

In between snippets of sleep, I reflect on the day's many blessings — the sacred mountain, the dances, the language, the Mudheads. How alive and vibrant are the Zuni people's spirit and culture! My mind drifts back to 1982, when I lived with the Ojibwe in northern Wisconsin as a Franciscan volunteer. My job was to direct a choir, teach music, and do whatever else my hosts wanted me to do.

One day a tribal elder, Mike Newago, told me his brother, Sticker, had died and he needed help digging the grave. I obliged and learned that pounding through six feet of stubborn red clay is a tough assignment. The Newago family thanked me with an invitation to join a group of Ojibwe youth and adults at a language camp. For two weeks we slept in tipis, built a birch-bark canoe, and gathered medicinal plants. We were allowed to speak only the Native tongue. For me, that meant living mostly in silence, though I did pick up enough words to identify many animals, some plants and, most important, whatever was being served for supper.

My most lasting impression came from an older Ojibwe woman who helped me make a small basket from birch bark and spruce roots. Thirty-five years later, I still have that basket. It's seen continuous use and sits on my desk holding keys, wallet, glasses, and other personal items.

The basket shows the wear and tear of time but still serves its purpose. Most significantly, its presence is a daily reminder of the sustainability of Indigenous culture. A basket made of plastic lacks both the durability and aesthetic appeal of one made from living beings. It also lacks any spiritual value, and even seems to mock those values. I think of all the ways in which a plastic basket destroys life —

extracting oil, refining oil, the factory where it's made polluting the air and water, the impoverished lives of the factory workers, the emissions from trucks and ships transporting the basket from a sweatshop in southeast Asia to some Walmart in Anywhere, USA.

No environmental degradation occurred in the production of my basket. The only lives lost were those of two noble trees: a birch and a spruce, whose raw material supplied enough bark and root to make many similarly durable products, whose hollowed trunks provided shelter to generations of mammals, whose decaying bodies fed countless bugs and birds, and whose life force gradually enriched the same soil that had nurtured them while alive.

Unlike most of what humanity tosses on the garbage heap of rampant consumption, if one day my basket ends up in a landfill, the remaining scraps of birch and spruce will prove an asset to the area's water and land, not a toxic menace to be covered with dirt and monitored for centuries to come.

I drift back to the here and now and consider that every moment contains its unique blessing. The toilets echoing through our sleeping quarters are a blessing, as are the accompanying spells of sleeplessness that provide a calm space in which to integrate the lessons learned from the Zuni and Ojibwe people.

A society built on the rapid, violent exploitation of Earth cannot endure. Modern society will either pass into oblivion or be supplanted by sustainable models of governance and economy guided by the morality and wisdom of Indigenous peoples. These are our two options.

I'm encouraged to see their Native cultures gaining momentum across the world. Their continued ascendency is one of two cornerstones that will save us in the New Climate Era.

The other cornerstone became a central component of One Earth Village.

CHAPTER 16
ONE EARTH VILLAGE

"If we could change ourselves, the tendencies in the world would also change. As a man changes his own nature, so does the attitude of the world change towards him …. We need not wait to see what others do."

— MAHATMA GANDHI

Our salvation in the New Climate Era will come from the unlikely marriage of Indigenous wisdom and Western technology. The same innovative spirit that spawned the Industrial Age and created our current predicament holds the key to a future in sync with nature. Nowhere is that innovative spirit more evident than with the rapid developments revolutionizing solar, wind, geothermal, biodiesel, and other renewable energy systems.

One of my goals at the outset of the March was to power One Earth Village with renewable technologies — both to let visitors see how these systems operate and to cut the March's carbon footprint. Part of our challenge is the ever-changing design of camp. Residing in One Earth Village means that your entire home is rearranged every 24 hours.

Each day, Sarah has to rebuild the community, navigating the quirks, parameters, and opportunities of each new campsite. No one, including Sarah, can ever be sure how it will all come together, nor how our renewable energy features will fit in. Over time, marchers get used to the reality of daily newness and come to care more about basic survival than which direction the kitchen truck faces.

Central to our power system is a solar collector I rented from Derek Nelson in Des Moines, signing a contract for $600 per month. It takes about 20 minutes to set up the collector each afternoon and another 20 to break it down each morning. The collector works better than I ever imagined, even on cloudy days. It charges 30 phones and computers at once, along with our power-hungry solar refrigerator. As the charging ports are almost always full, some chaos is created by the inevitable entanglement of cords and wires, leading Sarah on one occasion to observe, "The solar collector looks like a sow with a piglet hanging off every teet."

Every gathering of humans must, it seems, have one curmudgeon. One Earth Village's grumpy guy is Pablo Howard. Pablo doesn't march much but is nonetheless a great asset to our camp, given his extensive knowledge of electrical and mechanical systems. Pablo's pickup serves as the designated break truck, hauling water, snacks, and the Eco Commodes.

Pablo has a set of solar-powered batteries. "They'll work just as well as your solar collector," he insists. "Let's save $600 a month and avoid the hassle of moving our power source every day. Send that damn thing back to Des Moines."

I tell Pablo I have no idea if his batteries will work as well as the solar collector. "I'll take your word for it. But we can't just send the solar collector back to Des Moines."

"Ken's leaving the March this week, driving east," Pablo points out. "I can rig it so he'll have no trouble hauling the solar collector with his Prius."

"That doesn't matter," I say. "Derek and I have a contract."

"You can always break a contract," argues Pablo.

I bite my lip. That attitude is so repugnant to me. One's word means nothing to some people. Whether it's a contract between two people or a treaty between, say, the US government and a sovereign Indigenous nation, there are those who will break or ignore agreements simply out of greed or expediency.

I think about the times as a kid in Ireland when I would watch my uncle and a neighbor "sign" an agreement, usually regarding the sale of a sheep or cow. When the haggling over price was accomplished and the deal made, both men would spit in their right hand, look each other in the eye, and shake hands vigorously. That was it. No paperwork. No attorneys. Just a transaction sealed with a wad of spit, a firm handshake, and direct eye contact. To the best of my recollection, a contract thus struck was irrevocable. One's reputation was important in rural communities and this primitive ritual ensured contractual compliance.

"Pablo, if you want to use your batteries to power our stuff, that's fine," I say. "And if the March continues to grow, we'll probably need them and maybe even more power sources. But I'm not going back on my word. That solar collector's staying right here."

Beyond the moral value of keeping my word to Derek, standing up to Pablo proves a wise decision. He suddenly leaves the March. We agree to buy his solar batteries, which eventually will power our refrigerator, leaving the solar collector to run everything else.

The loss of Pablo's truck leaves Sarah in the lurch as she scrambles to figure out how to move the Eco Commodes. These environmentally friendly toilets use sawdust instead of the usual blue chemical laced with formaldehyde. Instead of creating toxic waste for a landfill, Eco Commodes produce "humanure" — a compostable mix with environmental and agricultural value. They also save thousands of dollars over what we would have paid for conventional porta-potties.

But the Eco Commodes come with three unique challenges. First, marchers and camp visitors have to break the habit of peeing and pooping in the same receptacle. The uric acid in urine is toxic, so commode users have to pee in a vat that's later emptied into a standard toilet. Visitors with other excretory business sit on the Eco Commode's throne and empty their bowels into a five-gallon bucket, which is then covered with sawdust.

The second challenge is finding sawdust. Every two weeks, Sarah has to hunt down a source, usually a woodworking shop, landscaping business, or cabinet maker. We never once run out, but we cut it close on several occasions.

The third challenge is emptying the commodes. Since this was deemed the most objectionable of tasks, I volunteered for the waste disposal team at the outset of the March. Given my experience digging graves in Wisconsin and Minnesota, I figured I was reasonably qualified.

In the Mojave and Sonoran deserts, the ground was so hard and rocky it was often difficult to dig a hole deep enough to cover the waste. When the March came to regions of the country blessed with better land and more rain, it would prove easy to find farmers willing to let us bury the waste in their fields. Sarah became adept at finding a site every ten days or so.

Other renewable features of One Earth Village are our parabolic solar cookers and solar ovens. The cookers are large, concave dishes with a "burner" in the middle to hold the pot or pan. Before the start of the March, we tested a cooker in Des Moines with great success. One sunny day with the mercury barely above 0°F, we brought water to a boil in no time at all. Unfortunately, because of the flimsy stand and large dish, the cookers prove unworkable in even a modest breeze. They're also difficult to assemble and disassemble every day. After a couple failed attempts, our chefs abandon them, and we sell or give them away as we move across the country.

The solar ovens work better. These are rectangular boxes that sit low to the ground. They function well even in the wind and are easy to pack and unpack. Marchers boil water and have made snickerdoodle cookies, brownies, lasagna, pineapple upside-down cake, and more. Of course, on days without sun or when we have to prepare meals in the dark (which is every breakfast and occasionally supper), they're useless, and most of our cooking ends up being accomplished with two standing gas stoves.

For the second half of the March, a vegetable-oil-powered pickup truck owned by Gavain U'Pritchard hauls the commodes. The raw product for our biodiesel comes from grease Gavain obtains from restaurants along the way.

Later, Rob Hach donates a mobile wind turbine which stays with us through Washington, DC. Gavain hauls it in his truck and sets it up most days when he arrives at camp. The turbine is impressive and an excellent conversation starter for visitors and local media. But it arrives too late to find its niche in One Earth Village's power structure, and in the end is largely decorative.

Possibly the most well-managed "room" on the March is the mid-size U-haul truck we use as our kitchen. Mary DeCamp takes charge and rules it like a benevolent tyrant. Having a qualified, responsible point person devoting a large chunk of her time to this task proves instrumental in its success.

The gear truck is our least well-managed feature. Much of the problem stems from constant battles over how best to organize its contents, and I regret not having a firm system in place before the start of the March. The gear-truck challenge is further complicated by marchers constantly bringing in new stuff — banners, art supplies, musical instruments, lawn chairs, bikes, enough medical supplies to treat a battered army, more socks than we have feet, fire wood, and even a large fire pit.

One particularly daunting gear-truck conflict ensues when Bob Cook begins to sleep in it. At night, Bob removes people's gear from one of the lower racks, lays out his mattress, and claims a shelf as his bunk. Steve, Shira, and I often set out before Bob is awake. We try to quietly load our gear onto the truck, working around Bob as best we can. One morning in the dark, I nearly kick over a piss pot Bob set on the floor next to his mattress. I'm angry as I consider what might have happened if the upended pot of urine had soaked marchers' bags and gear.

Like the camel's nose under the tent, Bob's occupation of the gear truck encourages other marchers to follow suit. Next, the truck becomes a convenient place to drink and smoke pot. Not only do we have to worry about knocking over piss pots, we have to be careful not to kick over half-finished cans of beer, too.

As much as I want to exercise my authority as March director and end this foolishness, it's a matter that falls under the jurisdiction of our elected Council. The issue becomes contentious enough that an all-camp meeting is called. The Council opts to submit the decision as to whether Bob should be allowed to sleep on the truck to a vote of all present. Steve and I cast the only "no" votes.

Ironically, a couple days later, Bob leaves the March because he isn't feeling well. By the time we get to Washington, DC, Bob will have left and returned to the March a total of six times. After his next return trip to camp, Bob's able to find indoor accommodations each night, and the gear truck thereafter remains free of sleeping marchers, piss pots, and half-empty cans of beer.

While the three vehicles needed to transport the basic stuff of our camp — the kitchen truck, the gear truck, and the Eco Commodes — are essential, there are usually far too many extraneous vehicles tagging along. Cars proliferate and become a crutch, a way to get out of walking. With ample options to catch a ride to the next town, some marchers choose to hang out in cafes while the rest walk.

Today, as the March heads toward Grants, New Mexico, I set out from camp with Lala. She hasn't marched every mile, but she's strong and walks most days.

"I'm impressed with your gait," I tell her as we stride along. "I've got long legs and a whole foot of height on you, but I've gotta work to keep up."

"I'm used to it," Lala explains. "Back home, I walk five miles every day at 15-minute pace. Been doing that for years."

"That's quick!" I note. "We're probably doing, what, 17-minute mile pace now? That feels fast enough to me, especially with a long day in the mountains."

I change the subject. "By the way, I wanted to thank you for calling Kelsey off back there in Arizona. You handled that diplomatically."

"Thanks," Lala says with a smile.

"You know, you'd be a good legislator with that kind of skill."

"No fucking way! I hate everything about politics."

We walk for a bit in silence. "So, what are you gonna do when Shira leaves?" Lala enquires.

Shira's boyfriend, Tony, will soon join the March. He'll stay for two weeks, then he and Shira will head back to Massachusetts. Shira and I haven't talked about it, but it's a given that she'll move out of my tent and share space with Tony.

"Oh, it's ok," I say. "I'll miss having her as a tent mate. Mostly, I'll miss walking with her when she leaves after Taos. But I've known all along that it's coming. Besides, the nights are warmer now. And sometimes she snores. I won't miss that!" I jest.

"So what do you think of Jimmy?" Lala asks.

"How do you mean?"

"Well, I kind of like him."

"Jimmy's great," I say. "Really, I'm not sure why he's single, he's such an all-around decent guy. I've known him for five years at least. He'd be quite a catch," I say as I look at Lala, wanting to make sure she knows that I offer that observation in a protective manner. "Whoever hooks up with Jimmy better treat him well. He deserves someone who doesn't mess with him, who truly loves him," I say, to drive home the point of my glance.

We walk on in silence for another few minutes and I wonder if Lala isn't baiting me. Is she trying to make me jealous by bringing up Jimmy? Now that Shira's boyfriend is about to arrive, maybe she sees an opening. I remember the back rubs early in the March. I recall the look on her face when I told her I was sharing a tent with Shira. I think about her responses to my piano playing and wonder if her interest in me isn't romantic.

We come to a sign that reads, "No Dumping." I think nothing of it, but Lala becomes animated. She sees an opportunity for a photo. "Take my picture in front of that sign," she orders, as she loosens her pants, drops them to her knees, and squats.

"No, no," I plead. "Tell me you really don't want to do this? Who the hell wants to see you, or anyone for that matter, pretending to shit in front of a 'No Dumping' sign?"

"Just take the picture," she insists.

"I will if you promise not to give me a photo credit," I say. "I don't want to be an accomplice to any crimes of bad taste, and this one qualifies in spades."

Shortly after that, we come to the break truck. I stop only briefly, then push on by myself, wanting to get work done in Grants before arriving at camp. I make a few calls as I walk and try to process my conversation with Lala. She's good, I think to myself. Clever, perhaps even crafty. Bad taste in photography, but a great marcher and a hard worker. An incredible cook, too, and she's been an ally for me when I've most needed one.

I can't say for sure whether Lala's overtures toward me are romantic. Best to put it out of my mind. My heart is totally with Grace, although over the past week our conversations have been shorter and less frequent. She again seems distant and eager to wrap up the call. I detect a quiet, unnamed tension.

While I have no interest in allowing anything romantic to happen between Lala and me, it's comforting to have her on the March, both as an inspiring walker and an ally when things get dicey.

As we arrive in Grants, I could not possibly have known how quickly and dramatically my relationship with Lala would evolve.

CHAPTER 17

WALKING STICK

"[B]efore the gate -- my walking stick's made a river of melting snow."

— KOBAYASHI ISSA

Much of our route across the Southwest includes stretches of historic Route 66. In New Mexico, we discover an even earlier Route 66 built in the late 1800s, long since reduced to a string of forgotten desert pathways pock-marked with washouts. For several days we enjoy traffic-free travel as we follow the old route around mountains, across the tops of mesas, and through dry stream beds.

Occasionally, we pass abandoned buildings that once provided travelers with lodging, meals, whiskey, and supplies. Today, we stop for a break at a hotel that last housed guests a hundred years ago. The roof has collapsed. Trees and shrubs reach skyward from where bed frames and dressers once stood. The bricks that formerly defined strong walls now lay in piles scattered about. This hotel didn't crumble because the structure was unsound but because the world changed and the owners were unable or unwilling to adapt.

I stare at one pile of bricks partially buried under a gnarled web of weeds and tree roots. I'm reminded of the collapse of my marriage with Kristin. Unlike this hotel, the bricks of our relationship didn't just fall out. In what now seems like an extended fit of madness, I yanked them out one at a time. By the fall of 2006, these acts of demolition had inflicted enough damage to the edifice of our marriage that it was poised to topple.

In November of that year, not knowing what else to do, I checked myself in to New Melleray Abbey near Dubuque. (Yeah, I know. You thought I was going to say, "sought counseling.") I spent two weeks singing and praying with monks, helping with farm chores, and walking through the forest. The experience didn't save my marriage, but it did result in a purchase that, years later, would become an essential element of my journey across America.

Most monasteries have a cottage industry. I once visited one that made peach liqueur and another that specialized in fruit cake. New Melleray's industry is coffins. The monks' beautiful forest of oak, walnut, cherry, and hickory provides the raw materials. On the last day of my retreat, I visited the gift store to admire the carefully crafted wooden boxes. I wanted some memento of my time at the monastery but wasn't yet in the market for a coffin. So I settled instead on a fine hickory walking stick.

Two years later, my dad became the proud owner of a New Melleray coffin. Ever since, I've felt connected with him through the forest that provided wood for both his need and mine.

My walking stick didn't see much use until 2013 when I began training for the March. I discovered that, if used properly, my arms would absorb some of the impact, providing my legs with notable relief. I'd practice planting the stick firmly in front of me, then I'd push hard until I could feel the tension in my triceps and forearms. As the stick completed its motion from front to back, there would be a slight lift in my stride and an increase in speed. To prevent the walking equivalent of white-knuckle driving, I'd roll the stick in my hand when I brought it forward, and I'd often switch arms. It took time, patience, and much practice, but I learned that a walking stick, used properly, is indeed a huge benefit to the long-distance walker.

After marching close to 1,000 miles through three states, my stick has become an essential companion. In addition to what I learned while training, I've become skilled at using it for balance, especially when walking through ravines or on sloping shoulders of gravel. In cities, I use it to push walk buttons, saving a few strides at road crossings. In rural areas, I use it to fling roadkill off highways.

My stick also becomes an important part of the story I share with people I meet. One day, Steve and I run out of water. "Look! There's a farmhouse ahead," I say. "First one we've seen in miles, and likely the last one we'll see before we die of thirst."

"Nah, let's just keep going," Steve says. "I don't want to bother the guy. We'll make it."

"Ok, you keep going and I'll throw dirt over your parched corpse when I find you up the road later today."

"Hello!" I shout as we get close to the barn. A wiry, weathered Latino man, a bit older than me, steps out. "We're walking across the country," I say, "and we've run out of water. Would you happen to have a bit to spare?"

"Sure," he says. "Walking across the country. What for?" he asks as he unkinks a hose to fill our bottles.

"We're trying to get people to understand that we have to do something about climate change," I say.

The farmer nods his head and asks, "How far do you walk in a day?"

"Oh, we probably average 16 miles or so," I say. "Today, we've got 18 to tackle."

The farmer looks at my walking stick and say, "That's a good hike. I'll bet that stick helps."

"It sure does!" I affirm. "It reminds me of my dad, who died of cancer a few years back. His coffin was made from the same forest that grew the wood for this stick."

The farmer seems genuinely interested, so I continue.

"My dad used to tell me what it was like growing up in the Bronx during World War II, the son of Irish immigrants. Military experts warned of the horrible things happening in Europe and the South Pacific. But our leaders just sat on their hands. Eventually, America woke up and everybody did their part. My dad would go down to the railroad tracks and collect tinfoil from cigarette boxes for the scrap metal drives."

"Yeah, my father fought in World War II. I was born a year after it ended. Not a coincidence I guess," the farmer says, smiling.

"My dad. Your dad. Everyone got involved," I say. "As a nation, we retooled an entire economy in just a matter of months. We defeated Fascism, and who knows what the world would look like today if we hadn't."

There's a pause in the conversation as the farmer hands back our water bottles. We thank him. "Well, these times aren't that different," I say. "It's time to wake up, like America did back then. We've gotta listen to the experts, to the scientists who warm us of climate change."

I pause, then continue. "That's why we walk, to wake people up to what's going on, before it's too late."

The farmer is thoughtful as he stares at the ground. He glances up, looks at my walking stick, shakes his head slowly, and says, "Makes sense to me."

That story — first articulated in a barn in New Mexico, inspired by my dad and a hickory walking stick made by monks — becomes my core message during the rest of the March. I share that message hundreds of times over the next 2,000 miles. It almost always resonates a reminder of the power of stories in communicating with people about a crisis as urgent and intangible as climate change.

The burden of walking while working — then working more when I get to camp, then working every "day off" — wears on me more and more. I'm in daily dialogue with coordinators in four states. They're doing great work, but they need guidance, direction, and reconnaissance from the March. There's also the ongoing challenge of raising money.

I share with Steve how tough it is dealing with the physical strain of walking plus the mental strain of organizing and, on top of that, the emotional strain of marcher conflicts.

"Have you had any blisters?" asks Steve.

"No, no yet."

"Has your back been buggin' you?"

"Nope."

"How are those sunburned lips treatin' ya?"

"Better."

"Then quit yer belly-achin'," says Steve.

I pause, and consider my reply. "Gandhi once said, 'the woes of a Mahatma are known to a Mahatma alone.'"

"Oh, so now you're comparing yourself to Gandhi," Steve laughs, shaking his head.

"Maybe you'd prefer the quote about walking a mile in someone else's moccasins," I quip. "Seriously, I don't think anybody on this freaking March knows what I do every day. It's tough enough to walk, but I arrive at camp and then the shit really hits the fan.

"Yeah, all teasing aside, I know you got a lot on your plate," admits Steve. "I know it takes a lot to keep this thing going."

"Marchers get to sit around the fire every night. They socialize, play music, drink wine and beer, smoke pot. Yeah, I know they smoke pot, some of them. Not you, you're too damn square for that."

I'm genuinely worked up now. "Every chance I get I'm raising money, Steve. I'm reaching out to the media. I'm trying to make sure we have the next state organized. Jimmy and some of the other so-called marchers, they sit in a cafe every day. They've got their Internet connection, good phone service, plenty of coffee and food."

I know the tension between Jimmy and me has been building and that, at some point, something's going to break.

"I was sure Jimmy would be a great marcher," I tell Steve. "I thought he'd walk and work, like I'm doing. He's young, strong. Heck, when we walked together in Des Moines, he could keep up while playing the damn fiddle. I don't get it. I'm disappointed."

I think about Jimmy, wonder what he does during the days. I've hardly seen him at all lately, not even in camp. But I did notice that he and Lala are now sharing a tent and appear to be romantically involved.

"At any rate, Steve, marching is the most important thing I can do in this cause. A close second, raising funds to keep it going. Third, putting up with your shit."

I glance at Steve to see how he'll respond to my rant. He doesn't say a thing, just keeps walking, looking straight ahead. I take that as a sign that the burden of wrapping up the conversation is on me.

"So, Steve. Stop being an inconsiderate asshole," I say smiling. Steve smiles back. We keep walking.

CHAPTER 18

ALBUQUERQUE

"{W}hen we act, when we publish, when we break the law we do not agree with, when we act politically, we empower ourselves...we have saved ourselves because we have done something to change ourselves, and ultimately, to change history."

— DANNY LYON, **BURN ZONE**

After a week of pleasant weather, the cold and wind return. The sky is at war with itself as we arrive at our campsite on the Laguna Pueblo reservation, a day's walk from Albuquerque. A dust cloud to our north threatens to clash with a thunder cloud to our south. One Earth Village is caught in the middle of this celestial drama — naked and inconsequential, of no concern to these two giants intent on destroying each other.

It's hard to discern which storm poses the greater risk. Which god do I hope prevails? The god who wields the dark, cold thunderbolts and threatens to drown us? Or the god who commands the swirling dust clouds that blind and choke? In Greek mythology, when forced to choose sides in a dispute between gods, the mortal always loses. Regardless, I root for the God of Thunder, reasoning that recovery from a drenching will be quicker and less painful than recovery from a sand-blasting.

Miraculously, we avoid a direct hit as the two storms collide just east of us. But with more bad weather forecast tonight, the Laguna people let us sleep in the nearby casino's conference room. We're once again grateful beneficiaries of Native kindness.

Conditions improve as we arrive on the outskirts of Albuquerque the next day. A radio station interviews me and I announce that our motley collective of minstrels — the Climate Justice Gypsy Band — will perform tonight at a local cafe. We hope for a decent crowd and maybe a hundred bucks in the tip jar.

I arrive late. The band is on its second set when I grab my guitar and join in. Marchers are having a great time, despite the sad reality that our audience consists of one guy. He's an older man with large tufts of white hair sticking out of a well-worn baseball cap. He seems sad, or at least serious, maybe even a bit grumpy. He wears a faded black sweatshirt reading "Queens." That and his New York accent suggest he's not native to New Mexico.

Our tip jar is empty. The guy notices and drops a $100 bill into it. As the evening wears on and our band continues to play to a mostly empty house, the man drops in another $100 bill, then three more until the jar contains $500!

I introduce myself during a break. The man's name is Danny Lyon. He heard my interview on the radio earlier today and wanted to meet us. I thank him profusely for his generosity. He gives me his card and invites me to visit his home tomorrow, about 20 miles away.

I do a little research and learn that Danny's a distinguished photographer, journalist, and film maker with award-winning work dating back to the 1960s civil rights struggle. He picks me up at camp in a battered old Volvo and we drive to the adobe house where he and his wife, Nancy, have lived for 38 years. "Most of this house was built by a single illegal Mexican worker named Eddie," Danny says proudly. "And I like that it's biodegradable. Someday, it'll just be a big heap of mud."

The home is modest. A large room serves as Danny's work studio. The walls are papered with his photographs. I admire the iconic images of the civil rights struggle, men on motorcycles, and the decay and demolition of Brooklyn, where Danny grew up. The photos span fifty years of struggle by average American men and women — sometimes fighting personal demons but more often the demons of institutional injustice, racism, and environmental destruction.

A MEMOIR FROM THE GREAT MARCH FOR CLIMATE ACTION

Danny's working on a book about climate change. He sees how it's affecting New Mexico through less snow-pack, windier conditions, and bigger and more frequent wildfires. "What you guys are doing is inspiring," he tells me. "It's important. Maybe that's hard to see when you're in the middle of it, but it's important."

I note the consistently serious nature of his work and the strong social commentary. I ask about the political statements he's trying to make. "This is powerful material," I observe. "You focus on stuff that's controversial, edgy. Have you had much pushback over the years?"

Danny ignores my question. "I don't see my work as political. If you become overtly political, you destroy the art in the artwork. And this whole notion of objective journalism — that's bullshit. You can't stand in front of a burning hospital that's just been bombed and not have an opinion. You're not a reporter. You're a person."

After talking for an hour over tea, we adjourn to two swings in front of the house where I again thank Danny for filling up our tip jar last night. I tell him it's been a huge struggle to keep the March going across deserts and over mountains. We're hurting for cash worse now than at any point during the March. I ask him if he'll consider another contribution. "I know it's a big ask, but $5,000 would get us to Colorado. Could you do that?"

Danny says, "Nancy and I never give that kind of money. But climate change — I wish there was a better term for it, like 'the end of life as we know it' — is so distressing, so extreme, so beyond the pale of anything we can imagine. It gives me hope to see what you and these other marchers are doing. I wish there were hundreds of you. So, yeah, out of sheer terror, I'll sell a painting and give you five grand."

Danny drives me back to camp, then spends a couple hours talking with marchers. He photographs some of them. He's fascinated by Sean Glenn, who marches in silence and hasn't spoken a word since Los Angeles. Sean often walks barefoot. "That's pretty extreme," says Danny. "It demonstrates a level of commitment that I admire, that we need. Sean's silence and the fact that she's shaved part of her head even though she's young and beautiful — those things appeal to me, help tell the story of why you're all marching and why fighting climate change is so urgent."

Danny photographs Ben Bushwick and Judy Todd having lunch under a railroad trestle. "That's what hobos do," he says. "A moment like that says volumes about the sacrifices you make to be on this march."

He offers constructive criticism, too. "I see you marching or at the rally or concert and you're all wearing those Climate March t-shirts. It's a nice shirt, but only a few of you need to wear them. When you all wear them, you look like a team, not a movement. Teams are fun to watch on the baseball field or basketball court. But it's hard to take a team seriously when you're trying to save the world."

More and more, my body feels the cumulative fatigue of long marches through rough terrain and challenging weather. Many marchers have suffered blisters, leg injuries, or back pain. The worst injury is Mack's broken toe, which has yet to stop him from walking every step of the way — on crutches! I'm concerned that he's doing permanent damage, but he refuses to quit. Mack is indeed an inspiration.

I've been lucky in comparison, with only cracked lips, diarrhea, and general fatigue to contend with. But I know I'm pushing my body beyond anything it's ever done. My carnivorous, 5,000-calories-a-day diet is helping. Without it, as Charles assured me, my marching days would have ended in Arizona. But I worry that the relentless physical exertion of walking 100 miles a week may lead to an injury that forces me to quit.

Today, there's plenty else on my mind and I'm grateful to have two days off from marching. Danny's generous support buys us some breathing room, but I need to push hard to avoid another cash crunch in Colorado. I keep hoping some unknown benefactor will step forward with a check for $50,000 or $100,000. To date, most of our funds have come from friends and supporters in Iowa. Most of them are tapped out. So today on our day off, I find a cafe, a restaurant, and a park in Albuquerque's historic town center and migrate from one to the other, reaching out by phone to new donors. The calls go well, and by the end of the day, we have enough cash to see us through the first two days in Colorado.

Tonight, the March Council calls an all-camp meeting. A handful of visitors and short-term marchers join us as I pull up one of the many camp chairs we've accumulated over the past two-and-half months. I pan the faces of other marchers. Like the dust and thunder clouds that amassed on our way into town two days ago, I sense that things are about to get ugly.

The meeting starts with Kelsey and Michael issuing an ultimatum: if I don't give up my salary, they'll leave the March tomorrow. Kelsey wants to fire all March staff, but her compromise position is to target just me. A few marchers try to talk them out of leaving. One argues in favor of paid professionals in any successful movement. Another praises the staff for the long hours and hard work that have helped us make it nearly a third of the way across the country.

Kelsey and Michael remain adamant. Either I step down or they leave. I say nothing, but am quietly relieved by Kelsey's pending departure.

When it's my turn to speak, I simply announce that, in addition to the $500 in tips at our concert, Danny has agreed to donate another $5,000. Also, my phone calls today have generated another $1,000. I share this news, then sit back down to scattered, subdued applause, never responding to Kelsey or Michael.

Jimmy speaks next. For weeks, I've sensed a quiet but growing animosity from him, and now it gushes out. "Ed, you're being selfish. You can't both march and be the director. Other people are working as hard as you are and they aren't being paid. If you're going to march you should resign as director and turn the management of the March over to the entire community."

I've been calm so far, but now I'm pissed off. It's especially upsetting to be attacked by Jimmy. Coming into the March, he was a close friend. Yet since the Sonoran Desert, a chasm has gradually opened between us, widening with every passing day. It's now clear that he and a few other marchers want to wrest control of the March away from me and other staff.

"Jimmy, the problem isn't that I march. The problem is that you *don't* march. Some people, particularly Mary and John J., work full-time as volunteers. Sarah couldn't accomplish the daily challenge of moving and setting up camp without them. You're doing important stuff too. But if I can walk every day and get my work done, so can you. Hopping in a car and getting a ride to a cafe to sit in front of your computer all day is not what I hired you to do."

My words only exacerbate the conflict. Lala is furious. She literally screams at me, and not only says I shouldn't be allowed to march but that we're wasting close to $20,000 a month on staff for jobs that could be done by marchers. She then accuses Shari and me of mismanaging March funds and says she's digging up proof that she'll soon present to the March Council.

I saw the tension building between Jimmy and me, but Lala's anger takes me completely by surprise. I try to regain my composure and calmly remind people that our finances are and always have been transparent. Yes, we live on the edge, as do many non-profits. But if marchers have any concerns or questions about finances that I can't answer, they can talk with Shari. She'll get back with them as quickly as she can even though she's already got a huge workload and more responsibilities than most marchers can imagine.

More marchers speak, some reiterating Lala's concerns, others defending staff and me. The meeting fizzles to an inconclusive ending, though Michael and Kelsey repeat their intent to leave tomorrow.

I'm embarrassed that visitors and new marchers had to witness this. If they'd come to the March believing us to be a wonderful community of people working harmoniously together, they came away from tonight's meeting with a completely different perspective.

The next day we head north toward Santa Fe. I walk with Steve and Shira. "Well, how's that for irony?" I say. "After one of the most successful weekends of fundraising yet, the disfunction within One Earth Village sinks to a new low. Honestly, I'm relieved that Kelsey's gone, and if a few more want to leave, so be it."

"That could happen," says Steve. "If the exodus continues, the March could be in big trouble. Even though she's gone, Kelsey beat the anti-Ed drum so hard you're either gonna have to quit marching and sit in a cafe all day with Jimmy or give up your salary and figure out how to keep going."

"I'm leaving in two weeks," Shira reminded me. "Judy's taking off soon, and so is Debaura."

We walk in silence for a bit, then Steve says, "Marchers knew this was driven by paid staff when they signed on. But that don't matter a hill of beans. Shira and I know you work hard, but most of them don't see it. Besides, money *is* really, really tight. Somethin's gotta give."

"Steve's right," says Shira. "I mean, I live on a tight budget so I'm used to it. But this is different. Marchers joined knowing there's paid staff but they didn't realize we'd be this strapped for money."

More silence, this time longer, pushing the limits of a comfortable pause. "Well, I did some thinking in the tent last night, when you were snoring and I couldn't sleep," I tease Shira. "When it comes to money, I've made two miscalculations. First, I should've known there'd be resentment about paid staff. Right or wrong, that's just human nature. And I should've known money would be tight. Honestly, given that we're marching for the survival of humanity, I thought more big donors from across the country would step forward. Maybe they will, but we can't count on it."

"Yeah, what Danny did for us this weekend was great," says Shira. "But that's not going to happen very often, if it happens again at all. You've gotta cut expenses, and staff is one place to do it. Sorry," she smiles.

"I gotta agree with Shira," admits Steve, who pauses, slows his stride a bit, and looks right at me. "But remember, the most important thing you or any of us can do on this march is to continue making these footsteps. The daily sacrifice of grinding through 15 to 20 miles, putting one foot in front of the other in spite of all the obstacles, that's the strongest statement we can make to the world about climate change. So whatever you do, even if you have to quit being paid and get by eating tofu and couscous and maybe the occasional squirrel, do it. Just keep marching. I'm counting on you, and I think most of these marchers are. Tune out the crazies and those who just want you to stop cause they ain't strong enough to do it themselves."

Steve continues to look me in the eye for a few seconds, then says, "There, that's my advice. Probably not very good advice, especially if you lose another 20 pounds and fall over dead goin' over the Rockies. But it's the best I can offer. And if you like, I'll teach you how to gut and cook a squirrel."

During the next mile, we walk mostly in silence as I process the conversation. If I could do it all over again, I would never have been paid by the March. I would have continued to earn a living through my talk show and worked on the March as a volunteer. But now I'm in a bind. I have bills to pay back home and my pricey visits to cafes are my only defense against auto-cannibalism.

"Ok, I'll tell you what," I announce to Steve and Shira. "I'm going to try to find a way to give up my salary and still walk. I'll talk with friends and supporters back in Iowa and maybe they'll help. But it's awkward. I don't have any trouble asking for donations for a cause or a group effort like this. But when that cause is me it's harder. But it seems like it's the only option. Besides, even though I'd eat squirrel, if Steve is cooking it, I bet it'd taste like shit."

CHAPTER 19

FARMER MILLER

> "There are two spiritual dangers in not owning a farm. One is the danger of supposing that breakfast comes from the grocery, and the other that heat comes from the furnace."
>
> — ALDO LEOPOLD, *A SAND COUNTY ALMANAC*

In the tiny village of La Bajada, Sarah lands us a beautiful campsite on a small, tidy farm. The strip of grass we call home tonight is a tight but comfortable fit, our tents and vehicles clustered between cropland and the Santa Fe River. The river here is narrow, shallow, and shaded with trees. Marchers sit or play in its cool waters as the dinner crew prepares a rare treat for the omnivores in our group — salmon, gifted to us by a local benefactor.

After setting up my tent and resting for a bit, I wander off to talk with the farmer. He introduces himself as "Miller" — I never do catch his first name. He's a decade or more younger than the average Iowa farmer, perhaps in his early 40s, sporting long hair tied back in a pony tail.

Farmer Miller gets off his tractor as I thank him for his kindness. He's friendly and grateful for what we're doing, but angry at the world.

"If things continue the way they're going, I won't be farming much longer," he warns. "There's less rain all the time, and now there's a water war between farmers and the City of Santa Fe, which Santa Fe is winning." Miller claims that a lot of Santa Fe's increased water consumption is for golf courses. I don't doubt that he's right.

Like nearly everyone we meet, Miller is curious about the March. He asks about our route tomorrow. I point to the towering cliff to our east, which bears the same name as the village — La Bajada, or "The Descent."

I explain that we'll follow the rugged switchbacks up the cliff, then trek another 18 miles across the desert until we hit Santa Fe. Miller stares at me for a few seconds then turns to look at the steep ascent above the village, perhaps pondering the long, barren expanse of land beyond. His eyes grow wide as he says, "That's a death march, ya know," hoping he might somehow talk us out of it.

"Well, it won't be our first death march, and probably not our last," I chuckle, trying to reassure myself at the same time.

Shira's friend, Tony Pisano, arrives today. He's a gentle man with a pleasant disposition and striking appearance. Short with a wide, expansive mustache, there's something about Tony that suggests a gnome. At age 60 he's spry and a strong walker. We're logging long days at high altitude. Steve, Shira, and I are impressed that Tony can keep up with us.

Tony plays the accordion. For the next two weeks my old squeeze box sees more action than any other time on the March. I joke that there's a law in New Mexico limiting the number of accordion players to one per group of 25 otherwise normal people. "So, Tony, one of us has to leave," I banter. "You wanna take over as March director? It's a blast. Everyone will love you. I promise."

Alone in my tent that night, I think about Miller. It saddens me that his days tilling the soil may be numbered. For nearly a century, small farmers like him have been pushed off the land by policies and priorities favoring large farms and development interests. As climate change accelerates, the exodus from farm country is likely to worsen.

Farming was my dream, too, but back injuries forced me to quit. In 1984, Kristin and I had moved to my family's farm in Ireland. We raised several head of cattle, a flock of chickens, a large patch of potatoes, and three fields of hay.

Farming in Ireland was just beginning its transition from manual to mechanized. My cousins, Joe and Kathleen Moffat, were at the tail end of that transition. They used bicycles to get around. Joe once showed me how to "fix" a flat by weaving a tight circle of straw and stuffing it into the tire.

Like the Moffats, Kristin and I were carless. We dug our garden with hand tools. We brought buckets to the well to haul potable water. Our lone source of fuel for cooking and heating was turf, and in late August we biked to the bog each morning to "make turf" until sundown. Lifting the heavy bags onto a cart drawn by a donkey proved more than my back could handle. One day, my back just snapped and I spent the next two months in bed.

Later that fall, Kristin and I moved to Iowa. My back was miserable for the next 10 years, but I still managed to satisfy my love of farming with a small garden that got bigger each year. Although we were poor and relied heavily on what we grew, I always took it in stride when one or more crops failed.

Except potatoes.

When the potato crop failed, instead of an "oh well" attitude, it shook me. Why, among the 25 to 30 different vegetables and fruits that we grew, did a disastrous potato harvest elicit such a reaction?

I thought about it a lot and concluded that the Great Hunger of the mid-1800s was to blame. During our year farming in Ireland, there was no blight, no risk of starvation, no shortage of food. Yet during the growing season, every farmer in the area sprayed the family's potato crop six times for blight! The traumatic imprint of a famine 140 years earlier remains branded into the Irish psyche. Despite the passage of time — for me, six generations removed from the famine and two from Ireland itself — I'm still wired to fear starvation if my potato crop fails.

There's a broader lesson in this truth, one that America's "dominant" race needs to understand. I've never heard anyone tell an Irish-American kid to get over the potato famine. Yet how often have I heard it said that Blacks need to get over slavery, or that Native Americans need to get over attempted genocide?

Privileged White Americans either don't understand or choose to ignore the fact that what happened in the past matters. If your great-great-grandfather was chained in a ship crossing the Atlantic, then humiliated and beaten for the rest of his life while legally defined as three-fifths human, do you really think his descendants won't bear those scars? If your great-great-grandmother was raped by soldiers as she watched her village burn then forced to march hundreds of miles through blazing heat or freezing cold to see her children die or be sent to reservation schools for indoctrination, is it that hard to understand that it takes many generations to recover from the horror of colonization?

There is one "get over it" that makes sense. White America needs to get over the denial of its privilege — even those of us whose ancestors immigrated from the impoverished backwaters of Europe to escape famine, war, and genocide. We whose ancestors were maligned by earlier White settlers when they arrived nonetheless have a distinct advantage over non-Europeans — skin color. While America has made great strides against the scourge of racism, we still have a long way to go.

Rich in cultural and historic appeal, Santa Fe is the perfect town for a day off. Yet, a ton of work demands my attention. So I head to a cafe to work. I pass several public pianos and want badly to stop and play. A block from the cafe in a small park I spot a multi-colored upright piano. As I draw closer I see a woman playing to an otherwise empty park. I hunger to listen but instead hurry along.

I reach the cafe and, as I grasp the door handle to enter, stomp my foot. "Damn it, Ed," I say out loud. "The cafe will still be here in 30 minutes. Work can wait. Go listen to the piano player, maybe play a few tunes yourself."

I arrive back at the park just as the woman is wrapping up a sweet, soulful improvisation. I thank her and she asks if I'd like to play.

"Sure, if you help me drag that picnic table over so I can use it as a piano bench," I say. "I don't know how you can play standing up like that."

The woman's name is Lacy Saunders and it takes us all of 30 seconds to rearrange the park's furniture. As I begin to play a Chopin nocturne, Lacy takes out her phone and asks with a nod if she can record. I nod back then lose myself in the music. When I finish I look up and see two new arrivals applauding with Lacy.

"That was beautiful," says Lacy. "Play another." I do, and the audience grows to a half-dozen listeners. More applause.

Lacy's a singer and we cautiously agree to take a crack at "All of Me." "It's been at least thirty years since I've played that tune," I confess. But I hold my own. Lacy's got a beautiful voice and by the time we finish, our audience has tripled. More applause.

"That was a blast, and look at the crowd we're drawing," I say just as a young man with a guitar shows up. His name's Will, and we invite him to join us. We play "Summertime." Will's quick, and he figures out the chords with no trouble at all. We then play "Georgia on My Mind." By now the crowd is nearly fifty people! "Wow, folks here must be starved for this kind of music. Think we can line up a gig tonight?" I jest.

As I pan the crowd I notice lots of dogs. "Yeah, weird," observes Lacy. "And every one of them is a German Shepherd. Wait, they're all wearing numbers. What the hell?"

At that moment, a man leans over the piano and says matter-of-factly, "Excuse me guys, we need to judge a Rin-Tin-Tin look-alike contest. Can you stop playing?"

"Sure," I say, after a slightly awkward pause. I smile at Will and Lacy and we just sit there for a bit. As the contest gets underway, we look at each other, smiles breaking into laughter. I'm tempted to punctuate the contest with appropriate little riffs and ditties. Perhaps two measures of "How Much is That Doggie in the Window," or a lick from Elvis Presley's "Hound Dog." But we all have places to go. Lacy's headed to a jam session. Will vanishes as quickly as he arrived. I've got March work to attend to.

I head back to the cafe and think how much richer my life is because I did what I wanted, not what I felt I had to do. Granted, always doing what one wants is the life of a child. Today, I'm grateful that I found the balance between duty and pleasure. What a rare blessing for three people who've never met to discover how beautifully their musical gifts blend — and then the high comedy of having that musical session interrupted by a herd of well-mannered German Shepherds! This is the kind of spontaneous, unscripted moment that's easily missed when rushing through life.

I've learned that pain, joy, and even mundane affairs are amplified through the raw existence of the March. The final leg of our journey through New Mexico was about to further demonstrate this truth.

CHAPTER 20

MARCHER

"A pilgrim is not a tourist who only touches, for a fleeting moment, the land and people that they visit. Rather, a pilgrim seeks to understand the essence of time, place and people that they meet on their path."

— SANTUARIO DE CHIMAYO WEBSITE, HOLYCHIMAYO.US

Our itinerary during the final push through New Mexico reads like a field trip for a course in comparative religions. On May 19, we pitch our tents at Santuario de Chimayo, one of the most prominent Catholic pilgrimage sites in the country. Most of the center's 300,000 annual pilgrims arrive by car or bus, but an impressive number make the journey on foot. There are annual pilgrimages of 100 miles from within New Mexico. A longer trek of 360 miles — Camino del Norte a Chimayo — starts in Denver. Presently, one woman is en route to Chimayo from Illinois — over 1,200 miles away!

Chimayo's website could just as well refer to climate marchers, describing pilgrims who "experienced hot, dusty, sweaty, confused, and exhausting days." Today, two apparently confused marchers bathe naked in the stream running by the main parking lot. When I find out about it I want to scream, "Skinny dipping? Really? At a Catholic retreat center kind enough to give us a place to stay? What the hell were you thinking!" Fortunately, the matter was addressed by calmer heads earlier in the day with no backlash from the center's staff.

A stone statue of a wandering pilgrim sits prominently in the heart of Chimayo. "Looks a lot like you," says Steve as he insists on a photo, directing me where to stand and how to pose. The statue and I both sport wide-brimmed hats, satchels, and walking sticks. In his free hand, the pilgrim holds rosary beads. Mine clutches a cell phone.

I admire the pilgrim's composure and purpose. The relative simplicity of his journey is enviable. Walk, pray, eat, sleep. "Yeah, I guess we look a lot alike," I tell Steve. "But he's a pilgrim. I'm a marcher. When you're out to save your own soul, a pilgrimage is great. When your mission is to save a planet, you march."

The next day, Steve, Shira, Tony, and I set out before sunrise. We skip breakfast on the promise of a cafe in Truchas — nine miles away and a 2,000-foot climb.

The ascent is difficult but beautiful, our path framed by snow-capped peaks. We pass a tiny hamlet nestled in an emerald valley far below. The intense quilted shades of green remind me of springtime in Iowa, and I'm drawn to the valley's lushness like a bug to a blue light.

We're beyond famished as we close in on Truchas. Steve has developed this amusing habit of announcing "I smell bacon!" when we approach a town. As we pass the first house on the edge of Truchas, he loudly declares, "I smell bacon," then glances at me, scowls a bit, and says, "There'd *better* be bacon."

"Yeah, yeah," I reply. "It's a big town, bigger than Chimayo, which had a great restaurant. I promise you bacon — bacon and more."

At the next house, a woman is just stepping out to the edge of the road to put up a sign advertising her art gallery. "Excuse me, ma'am," I enquire. "Can you tell us how to get to the cafe?"

She laughs and says, "Sorry, but there's no cafe in Truchas." Steve sighs a long, sad sigh that only a man who's sustained himself with visions of a hearty breakfast while trudging for three hours up a mountain can sigh. Undaunted, I ask "I guess we'll have to settle for the convenience store. Can you tell us where that is?"

"We don't have a convenience store either," she replies. Steve's head droops, hat covering his eyes, and I fear he'll collapse and die on the spot, but not before strangling me.

"Tell you what," the woman says. "Come in and I'll make you breakfast."

The woman's name is Trish Booth. She and her husband, Leonardo, live behind their art gallery. They offer us stools around a large wooden table in the middle of the kitchen. Trish prepares a hearty breakfast of eggs, toast, hash browns, coffee, and tea — everything but bacon. We eat as politely as our ravenous appetites allow and discuss politics, climate, art, and human kindness.

This is one of those unanticipated moments when I'm reminded of the essential goodness of people. After an hour of food and conversation, we visit the art gallery before striking out on the final six miles, mostly downhill. Trish and Leonardo later round up some friends and join us for supper and music at our campsite in the Santa Fe National Forest. This, too, is why we march.

The following day we camp at Southern Methodist University Retreat Center, where I witness the most astounding display of humming birds. Twenty or thirty hummers hover around a half-dozen feeders. I walk right up to them and they're not the least bit intimidated, buzzing just inches from my face.

As I often do, I pitch my tent away from the main encampment, finding a beautiful spot next to a lively stream that promises to sing me to sleep. Around 10:00, I crawl into my sleeping bag and check the forecast. I'm disheartened by warnings of heavy thunderstorms and flash flooding. It's too late to relocate my tent in the pitch-black mountain night, so I pack my gear and place it next to the tent door where I intend to grab it and flee if the waters suddenly rise.

They don't, but my mind is on high alert all night. Sleep comes intermittently. Several times I awake from dreams where I'm holding my phone in an outstretched arm, struggling to keep it and my head from going under as a cold torrent of water rushes into my tent.

The next morning, we march into Taos in a downpour. A new marcher joins us — Faith Meckely. Faith is fighting to protect Seneca Lake in upstate New York from fracking. She plans to walk the rest of the way to Washington, DC. Dozens of locals join us as well. At our rally, Lawrence Lujan — Lieutenant Governor of the Taos Pueblo — blesses us and Woman Stands Shining of the Diné tribe speaks about the urgency of our mission. Faith writes in her blog: "{Woman Stands Shining} told us how people on pilgrimage journeys always bring rain with them. She said we are pilgrims, and that our walk is a prayer, so it's no surprise we brought the rain."

All across New Mexico, Native peoples have welcomed us, guided us, fed us, sheltered us, and thanked us for what we're doing. Woman Stands Shining is not the first Native leader to call us pilgrims, to credit us with bringing rain to a region suffering intense drought and unprecedented wildfires. Did we bring rain to Taos? On the very first day of the March, were the heavy rains that ended Southern California's prolonged drought a result of our efforts?

My rational mind refuses to believe there's a connection. Yet this I know: whether we're pilgrims or marchers, whether we bring rain or are simply coincidental travelers caught in the storm, a whole lot of people are thinking and talking about climate change because of our efforts. This, more than anything, is why we march.

In Taos we stay at the Hindu Hanuman Temple where a flat, open field provides an ideal campsite. But the marchers who unload the gear truck leave much of our equipment sitting out in the rain. Other marchers arrive to find tents, sleeping bags, and clothing sopping wet. My accordion is among the victims. I'm incensed, yell at Sarah and anyone else I think might have been responsible, and storm off. I'm losing my temper way too often. Later, I find Sarah and apologize.

Meanwhile, Tony patiently disassembles my accordion and dries it in front of a wood stove inside the Temple. Marchers with wet gear are given home stays. Tomorrow is a rest day and we'll have ample time to dry our gear. Many marchers attend the prayer services at the

Temple. I'm too busy being pissed off and feeling oppressed by a heavy workload. In my world as both marcher and director, a "day off" is about as real as the Easter Bunny.

On our way north from Taos, we stop at the New Buffalo Center, a community founded over 40 years ago by hippies. Many still live here even as they near the last decade of their counter-cultural lives. In one of the kindest, most humbling gestures we receive during the March, New Buffalo residents wash and massage our feet. After a bountiful lunch we gather in the Center's sanctuary where each marcher is blessed during a solemn New Age ceremony.

We stay that night at Taos Goji Eco Lodge, owned by Eric and Elizabeth Vom Dorp. After supper, as the sun retreats into the western sky, I walk down the road to thank our hosts. I find Eric corralling sheep and goats into the barn and chickens into the coop. We talk as he finishes his evening chores. "We're the only large-scale goji berry farm in the US," explains Eric. "But we're also a pretty famous tourist destination."

I'm fascinated by innovative farming operations. Eric obliges me with a quick Goji Berry 101 tutorial, but he's more eager to tell me about the lodging side of their business. "Some of these cabins were built over a hundred years ago, and they've seen some pretty famous people. Aldous Huxley came here to write. In his spare time, he built an outhouse that still serves its purpose."

"Wow!" I respond. "I wonder what Huxley would have thought of our Eco Commodes." Eric and Elizabeth have given Sarah permission to bury our humanure on their farm. I want to make a quip about Huxley's shit and our shit nurturing the same trees, but opt for discretion.

"D.H. Lawrence. Elizabeth Kubler-Ross, and Ram Daas all spent time here, too," continues Eric.

"Good to know," I say. "If I ever decide to write a book about this March and need to escape the din of Des Moines, I know where to come."

Walking back to camp on a night so dark I could barely make out my hand at arms length, I hear harp music coming from our big tent. I'm sure it's a recording, yet as I draw closer I realize it's live. I squeeze into the crowded tent where a man in his thirties plays a small harp. His wife and two children sit next to him. A lantern bathes the scene in a primal glow, illuminating the unwashed faces of marchers crammed tightly together, all very intent on the music. Of all the spiritual encounters we've had this past week in New Mexico, this moment feels the most profound.

As I soak in the ambiance, I feel as if we're a camp of gypsies and a traveling minstrel has arrived to barter music for food, trinkets, and a place to sleep. There's a rawness to the man's music. I'm reminded that the harp was inspired by the bow — a weapon to kill food to nurture the body, transformed into an instrument to make music to nurture the soul.

I think of my life with Kristin, our two kids and our extended family of housemates, friends, and neighbors. At night, I'd work in my office as Kristin practiced her harp in the next room over. Like so many gifts that enter our lives, over time I took this blessing for granted, forgot its value, downplayed its importance. Sitting here in a tent filled with sound, light, sweat, and dust, I vow to more fully appreciate the gifts in my life — especially Grace. I vow never to take her for granted. As the harper plucks the chords of his final song, I close my eyes and give thanks that, in 1,000 miles, I'll be in my soul mate's arms for a brief time when the March comes through Iowa in August.

As we near the Colorado border, our tour of spiritual communities lands us at Kagyu Milo Guru Stupa, a Buddhist shrine where each marcher has an individual camp site — an odd diversion for a clan accustomed to lodgings clustered close together.

The next day we visit the Lama Foundation — an ecumenical spiritual center high in the mountains north of Taos. Eighteen years ago a wildfire ravaged the forest surrounding the center. The flames were headed straight for the community's adobe and wooden buildings, then mysteriously split and went around. The forest on either side lay decimated but the Lama Foundation was unscathed.

Our walk through New Mexico — interacting with Native, Catholic, Methodist, Hindu, New Age, and Buddhist communities — has been exceptional. While most of America remains blind to the changing reality of Earth's climate, here we enjoyed the company and support of people who truly comprehend the peril of climate change. They grasp the apocalyptic nature of the crisis that compels us to march, and they responded with profound kindness and appreciation.

I wonder if we'll be as well received in Colorado. I call my friend, Al Brody. Al's a long-time resident of Colorado Springs and is helping with logistics for our rally there. "Be careful in Colorado," Al warns. "They'll run you over here. Carry a flag. They won't run over the flag."

I take Al's advice to heart and attach a small American flag to my walking stick. Each day, we draw closer to the Sangre de Cristo Mountains — 14,000-foot giants that tower to our north, roughly marking our entry into southern Colorado. I hope Al's wrong. I want to believe that Colorado will be as welcoming as New Mexico. We'll see.

CHAPTER 21
FATHER AND SON

> "[The Russians] are a scurvy race and simply savages. We could beat the hell out of them."
>
> — GENERAL GEORGE S. PATTON

I was never at peace with my dad over the issue of war. He worked for the Department of Defense in the "shipping and handling" division. He never talked about what he shipped and handled, except when he helped transport a 59-foot Douglas fir across the country to replace the aging mast of the USS Constitution. Otherwise, the details of his work remained a mystery.

"What's he do?" I once asked a high school friend. "Load bombs on boats bound for southeast Asia?"

War terrified me from an early age. I first learned about Vietnam when I was six. I'd lie in bed paralyzed by fear at the roar of jet engines being tested at General Electric two miles away. I thought the Vietnamese were coming to kill us. Mom tried to reassure me that there was nothing to worry about. The noise came from the factory where she and my dad used to work, where they'd met in fact. It was a good place, nothing to be scared of. She made no mention of the napalm being dropped that night on sleeping Vietnamese children by planes powered by those very engines. Like most Americans in the early 1960s, Mom dutifully swallowed the anti-communist propaganda, ignoring the occasional images of truth and horror that slipped through the pro-war ideologues' shield of deceit.

A MEMOIR FROM THE GREAT MARCH FOR CLIMATE ACTION

Growing up, I don't recall ever discussing war with my parents. But Dad's pride in all things military was regularly on display. It was a given that my two brothers and I would one day serve in the US Army. At age nine, I beamed with pride as Dad stood me in front of him, admired my straight shoulders, and said how sharp I'd look when standing at attention in my Army uniform.

Dad had a limited selection of records, most of them collections of tunes glorifying military conquest: "Battle Hymn of the Republic," "The Caisson Song," "Battle of New Orleans," Tchaikovsky's "1812 Overture," that sort of thing. He played them so often I could sing or hum along with most of the music.

When I was 12, the movie *Patton* came out. Dad and I went to see it in what felt like a rite of passage in preparation for my inevitable entry into the armed forces. The Vietnam War was growing unpopular with more and more Americans and *Patton* was part of the propagandists' effort to shore up support. Mostly, the film depicts the glories of armed combat, though Patton's belligerence was impossible to conceal. Dad came away praising a brilliant military strategist. I felt like I'd just watched a demented madman orchestrate three hours of carnage and slaughter.

If the intention of this father-son outing had been to nurture my enthusiasm for joining the Army, it achieved the opposite result.

As my 18th birthday and the requirement to register for military service approached, I considered two options: conscientious objection or moving to Ireland. Remarkably, on the very day that I turned 18, mandatory registration was eliminated.

Two days into Colorado, at the base of the Sangre de Cristo Mountains, we camp at Fort Garland, established in 1858 to protect White settlers from the Utes. Like most frontier stockades, Fort Garland's functional life was brief, about 25 years — just long enough to subjugate the Indigenous population. The fort is now a museum. It would also be the only military facility to provide a campsite for us during the March.

We pitch our tents on the parade grounds where troops once

gathered before heading out to complete the forcible theft of the Utes' land. On our rest day, I take time off from March work to tour the grounds of the fort. I talk with museum director, Anita McDaniel. At a time in US history when Native culture and spirituality are on the rise, I'm curious how she feels about the fort's role in conquering the Utes.

"The fort's history is important because it tells the story of western expansion," Anita says. "It's a rich history, and we're developing interpretation which brings Indigenous and Hispanic points of view into the story — not just the European/Anglo American version. Since Fort Garland never saw any actual conflict, our story isn't so much about war but about people."

I ask Anita if Anglo America is doing a better job with Native relations today. "I'd like to think we are," she replies. "One thing I wanted to do when I became director was to see the fort become a place of reconciliation.

Yes, America needs reconciliation — at both the cultural and personal level. I think back to my relationship with Dad. Here was a career military man, a sergeant in the Army who went on to work in the private sector and then with the Department of Defense. Here was his son, detesting war and working in the peace movement, perhaps subconsciously trying to amend for the perceived damage done by his father.

Over time, some reconciliation between us did happen. But it was Dad who initiated it, and in ways I didn't expect. In my early twenties, I became a vegetarian. I thought Dad might malign my decision. Instead, he experimented with being a vegetarian for a week. When I began to work in the peace movement, I thought he'd be critical. Instead, Dad was supportive, insisting that he and I were both working for peace in our own way.

At age 61, Dad was diagnosed with multiple myeloma, a cancer which very few survive. The cancer stemmed from exposure to radiation while stationed in California. He and other soldiers were instructed to "play games" with the massive radar device under their charge. Soldiers would stand in front of the radar and try to jam it with their bodies, simulating Russian planes attacking the West coast. I could only imagine how much radiation soldiers were subjected to in this "game."

Dad was never bitter at the US Army for causing the cancer that killed him. But I was, and I tried for years working with Iowa Senator

Tom Harkin to get information. I wanted to know what happened to the other men stationed with Dad. I wanted to know if the US Army consciously exposed soldiers to deadly doses of radiation. I wanted to know how many other American soldiers suffered similar fates.

Despite my persistence and the involvement of a US senator, the Army effectively stonewalled us. In the end, they gave Dad $117 a month compensation for hearing loss. Dad was satisfied. He felt our concerns had at least been heard and the money was appreciated. I felt like a murderer had just declared his innocence against overwhelming evidence to the contrary.

Dad fought hard against cancer, living eight years longer than doctors said he would. Toward the end, there wasn't much left to him. He'd shrunk by over a foot and dropped 75 pounds. Through it all, he never lost his sense of humor, nor his willingness to seek understanding and reconciliation.

In the last year of his life, my transgender niece came to Dad to explain that she was no longer his grandson, but his granddaughter. My niece showed great courage in doing that, given that she was treated horribly by her father and some of her male cousins. She must have been terrified at how my dad would respond. Yet all he said was, "It doesn't matter. Either way, I'm still your grandfather."

During our two days at Fort Garland, surrounded by cannons, guns and all manner of war memorabilia, I think about Dad. I reflect on some of the hardest days of the March — the deluge in Los Angeles, crossing the Mojave, the blizzard on the high plateau of Arizona. The realization comes to me that Dad has been with me the whole time, present in the wood from the living forest that created my walking stick and his coffin. Yes, he's been with me on this march, this great adventure, this self-sacrifice for a life-and-death cause. He's pushing me forward with every stride, encouraging me to do something big, something important.

One hundred forty years after the soldiers at Fort Garland oversaw the theft of the Ute people's land and the attempted destruction of their culture and heritage, Anita, the museum's director, works to build bridges between cultures that once clashed. Today, at a former military stockade on the edge of the Rocky Mountains, I feel that my journey toward reconciliation with Dad is finally complete.

Here at Fort Garland, I cry for my father, for a life cut short, for a life I never got to know as well as I should have. I cry for the many times I failed to embrace his attempts at reconciliation. If I could pick one day of this 246-day march where he could join me in the flesh, it would be tomorrow, when we set out from Fort Garland and plunge into the 14,000-foot peaks to our north. As we muster on the parade grounds, I would pose with him for a photo in front of the Fort. I would let him use my walking stick, let him feel its power captured within those grains of wood tendered to perfection by monks whose every act was a prayer.

I would tell my father that, even though my disdain for war is as strong now as ever, I'm proud of what he did in the service of our country. I respect his passion for peace, the same peace I fought for as an activist. I'd thank him for being a fighter and commend him for the dignity and composure he maintained as he battled cancer for so long, way too long, longer than I would ever have had the courage to endure.

A MEMOIR FROM THE GREAT MARCH FOR CLIMATE ACTION

CHAPTER 22

ROADKILL

"Despicable creatures, vultures: without a doubt the most disgusting birds ever. I suppose they served their purpose, but did they have to be so greasy and ugly? Couldn't we have cute fuzzy rabbits that cleaned up roadkill instead?"

— RICK RIORDAN, **THE THRONE OF FIRE**

Leaving Fort Garland, we ascend into the Sangre de Cristo Mountains following Highway 160. It's a busy road littered with roadkill, most notably elk. In less than ten miles, I count eight of these huge, majestic beasts. We arrive at our campsite to find a ninth elk has collapsed and died right where we're supposed to pitch our tents. The putrid smell of rotting flesh is pervasive, and Sarah scrambles to find a new campsite farther up the road.

You learn a lot about roadkill when you walk America's highways. There's often a local flavor to it. Today is dead elk day. During two days in Arizona the predominant victims were horny toads. On other days in the desert lizards were the roadkill du jour. Later this month in eastern Colorado it will be box turtles. On a gravel road in Nebraska I'll pass three dead red-headed woodpeckers in a two-mile stretch. And in Iowa, we'll walk a backroad slick with more flattened frogs than we can count.

Mostly though, roadkill is a smorgasbord rather than a featured special — astounding both in terms of variety and volume. It's impossible to say just how many dead animals we pass during 3,100 miles of walking. I conduct an unscientific survey on three occasions and, including smaller victims like lizards, frogs, dragonflies, and butterflies, I figure I'll have viewed over 20,000 corpses by the time we reach the White House.

That's a lot of carnage to experience in eight months. At first, my reaction is disgust. But as the miles roll by and the body count mounts, my mind and gut can't process so much death. Disgust gives way to sadness and eventually to numbness — a similar trajectory to how Americans have adapted to mass shootings. At some point, the carnage becomes too overwhelming and you simply shut down.

During the first months of the March my response is to fling small- and medium-sized victims off the road with my walking stick. The grassy ditch strikes me as a more respectful resting place for the dead — and safer for animals who come to nourish themselves on the remains. My walking stick is strong and can hurl even a medium-sized raccoon off the road. As my competence at "flingage" increases, I barely miss a stride. On one occasion, Jeffrey and I come to a newly killed skunk. I dutifully send the mangled creature flying into the ditch. That proves to be a mistake. My stick is now heavily tainted with the essence of skunk, and Jeffrey heads off on his own while I walk alone.

Eventually, numbness sets in for me, too. I know I'm walking by a dead animal but it ceases to shock me. I no longer care and pretend not to notice. Like sleeping, eating, and peeing, stepping over close to a hundred corpses each day simply becomes part of the routine.

In Ohio, when I realize indifference is now my norm, I'm disturbed at this loss of empathy. I force myself to stop, look at the victim, examine it closely. Usually the animal's fur is matted in blood. Here's a fox, tongue hanging out, entrails spilled onto the pavement. Here's a possum, still in one piece but with its jaw dislocated, mouth agape displaying worn teeth that have gnawed through many a meal in their day. Here's a woodchuck, intact save for the pool of blood encircling its head like a halo, as if marking it for sainthood. Here's an unidentifiable mass of hair and skin that's been run over so many times all I can say for sure is it was once a mammal.

Viewing a recently killed deer, I marvel at the feeding frenzy as a thick, swirling cluster of maggots diligently works its way through the venison's prime cuts, as if choreographing a dance — thousands of maggots dancing, crammed into a mosh pit of deer flesh, ecstatically eating and twirling their way around this carcass-turned-ballroom in an orgy of survival and propagation. For maggots, America's highways are truly a blessing.

In Pennsylvania, I shoot photos of some of these dead animals and half-seriously consider publishing a Climate March Roadkill Calendar. Maybe I'll personalize the victims, give them names and write obituaries, try to create empathy among my audience.

"Billy Raccoon, struck dead by a Hummer. Billy leaves behind a wife, fourteen kids and twenty-two grandkids. He will be greatly missed by his good friends at the Dumpster Club."

"Millie Monarch, rendered a hood ornament at 60 miles an hour. Too young, she had just tasted her first milkweed and was thrilled about her upcoming vacation to Mexico."

"Bucky Whitetail, crushed by a semi. Preceded in death by his parents and two children, Bucky was skilled at avoiding hunters but never fully grasped the risks of grazing along highways."

Besides humans and maggots, another creature that loves highways is the turkey vulture. This ghastly looking bird is everywhere we walk, in every state, along nearly every highway and byway. Sometimes vultures circle over our heads, perhaps anticipating the possibility that a marcher will end up roadkill.

Why do humans find them so ugly? Perhaps because of their function as much as their appearance, performing a job so unsavory it doesn't even exist in our lexicon. Given the estimated one million animals[1] killed each day along America's four million miles of public roads,[2] it's no surprise that turkey vulture populations have increased annually at a rate of 1.79 percent since 1990.[3] Remarkably, between Los Angeles and Washington, DC, I never see a single road-killed vulture.

Many aspects of the life of a coast-to-coast marcher are unique, among them experiencing close-up so many dead and mangled

---

[1] http://www.culturechange.org/issue8/roadkill.ht

[2] https://www.fhwa.dot.gov/policyinformation/pubs/hf/pl11028/chapter1.cfm

[3] https://www.nps.gov/ever/planyourvisit/upload/LearnMoreReport.pdf

animals. I examine, photograph, and describe in detail some of these creatures not to be morbid but to confront the reality that modern civilization exacts a high price on any life that strays into the path of its "progress." We look away from roadkill because to pay attention not only compels us to confront the raw agony of death but to examine our personal choices involving mobility, lifestyle, and so much of what underlies modern living.

Various government and industry publications often remind us of the downside of cars and trucks — human fatalities, air pollution, carbon emissions, high insurance rates, urban sprawl. But rarely do we discuss the billions of non-human victims our car-centric culture destroys each year. It's easier simply to ignore roadkill, drive by it quickly, and let the vultures clean up the mess even as we try to ignore them, too.

For close to 100 miles, the breathtaking peaks of the Sangre de Cristo Mountains guide us northward. We've experienced incredible natural beauty throughout the March, yet nothing compares with the hypnotic lure of these snow-covered giants. We spend an entire week in their presence. I feel sorry for the drivers who rush by in a matter of hours — like the contrast between a light rainfall that barely moistens the soil's surface and a prolonged rain that soaks deep into the Earth. Drinking heartily of the sacred power of the Sangre de Cristo, I feel my mind and spirit absorb a balm they desperately need.

It's hard to take my eyes off the mountains. But ever mindful of the risk of a sprained ankle, my eyes shift constantly from mountains to pavement, panning the surface ahead, watching for the rock or divot that could put a quick end to my marching days.

My eyes also pan the road for treasure — or as it's more commonly called, trash. Several marchers have become skilled at harvesting the roadside debris of their choice. Lee Stewart collects spoons and has salvaged literally dozens for use in our kitchen. Marie collects license plates, hoping to find at least one from each state we walk through. Lala has dragged into camp an assortment of functional items: a wash basin, blankets, towels, tent pegs, and various pots and pans.

My roadside treasure hunt focuses on coins. I've already claimed $20, and by the end of the March will have collected $55. Steve has found only a few coins and is envious of my success. "You find all that money 'cause you walk on the left and most coins end up on the shoulder. I'm walking there today. You move over here," he orders me gruffly.

"You don't find anything because you're a half-blind, lumbering oaf," I tease. "But fair enough. You get the shoulder today," I say as we switch places.

After a few miles and not a penny richer, Steve loses interest. Besides, he's more comfortable walking on the pavement than on the gravel shoulder. We switch back, and less than a minute later I spot a dollar bill in the ditch. I grab it and say, "Ha! I told you you're a lumbering oaf," as I shake the bill in his face.

"You're lucky that isn't a $20, or we'd be wrestling in the ditch over it right now," says Steve. "You can keep your little one-dollar bill, but I imagine you'll gloat the rest of the day."

"Probably for the rest of the March," I correct him.

Safety and traffic haven't been as much of a concern in Colorado as Al Brody warned. Perhaps the small American flag I've attached to my walking stick helps, as does the grandiose style of waving I've perfected during the course of three months and twelve-hundred miles.

My "papal wave," as I call it, works like this: as a vehicle approaches, I flip my walking stick into my left hand. When the car or truck is 100 yards away, I smile broadly as my right arm moves slowly from my left hip up and over my head where my open palm begins to rotate back and forth until the arm reaches its full extension about 18 inches into the roadway. This big, sweeping wave has a bit of magic to it — cars and trucks almost always move out and away from us.

Perhaps the wave is effective in part because it's confusing. I try to imagine what drivers think: "Is that guy just being friendly, or is he in trouble? Is he dangerous or just some weirdo? Wait, is that the Pope? Hey look, kids, it's Forrest Gump."

Whatever goes on in the minds of drivers, the wave is highly effective at making an instant connection that improves our safety. Lots of drivers wave back. Only rarely do I get flipped off.

We always walk on the left, facing oncoming traffic. But lately the arc of the roadbed has been more pronounced. Steve and I suspect this might explain the mild but nagging pain in our left legs. With other marchers well behind us and little traffic today, we decide to switch to the right side of the road.

But walking on the right I feel naked and vulnerable. My wave is meaningless. Even with minimal traffic, I'm unnerved every time a vehicle approaches from behind. After a couple miles, we switch back to the left.

Moments later, as we're climbing a hill, a truck passing us from behind blows out a back-right tire. There's a loud explosion as tire shrapnel flies through the air. A piece hits Steve in the head. He's stunned but not injured. Most of the shrapnel fires off to the right — exactly where we would have been walking had we not just switched back to the left. Being hit by that volume of tire at that speed could have done some real damage or even killed one or both of us. Maybe the turkey vultures circling overhead know more than I give them credit for.

We keep walking. "That was a close call," I say as Steve and I pass another chunk of roadkill. "You scared of death, Steve?"

"I'm not looking forward to it," he muses. "But I suppose it's just something else to deal with, another step in my life. I don't relish it coming, but I certainly anticipate it."

"Maybe I'm weird, but I'm not really afraid of death," I say. "I see it as that long-awaited mystery vacation. You know it's coming, but you don't know when. You know the ride is gonna be uncomfortable, but where you're going and what you'll do when you get there are known only to that Great Travel Agent in the Sky. To me, that seems kind of exciting."

We pause a bit, each reflecting on the subject in our own minds. I say, "Everybody's got their belief about what happens after death. But if we're honest, we'll admit that we don't have a frickin' clue. I just hope we make it through this March without anyone getting killed."

"Well, we sure tried today," laughs Steve.

"Yeah. I know," I say. "Let's not do that again please. That's my biggest fear, you know, that someone's gonna get hit or even killed. I'd feel responsible if that happened, like it was my fault, since this damn March was my idea."

My concern about safety is warranted. Some of the younger marchers have become almost cavalier in their attitude toward traffic, even refusing to wear safety vests. On top of this, some of the earlier conflicts with marchers that I thought had been settled are about to resurface in ways that further isolate me from the community — and compel Steve to take an unexpected course of action.

CHAPTER 23
STORMS ON THE PRAIRIE

"The question is always asked by the curious travelers who have crossed the Plains at Interstate speeds, 'How can you live here without the mountains, the ocean, the woods?' But what they are really speaking to is their desire to 'get it' right away. The sublime of this place that we call the prairie is one of patience and looking."

— KEITH JACOBSHAGEN, *THE CHANGING PRAIRIE*

On June 7, we reach the highest point of our March, crossing the Front Range at 9,420 feet. Since we usually change elevation at a gradual pace, altitude sickness hasn't been a problem for most marchers. Our biggest climb in one day was the 3,000-foot ascent of the Mogollon Rim. Today the elevation change is again 3,000 feet — all downhill as we plunge toward the western edge of the Great Plains.

Our trek through Colorado has gone well. We've had perfect weather and a mostly quiet route through the mountains. Without the disruption of traffic and trains, sleep has been more restful. Marchers seem less physically and emotionally stressed. The March community is more harmonious than at any point since the Mojave Desert.

Forgoing my salary has helped, too, diffusing the hostility directed toward me in particular and toward staff in general. Jimmy follows suit and now works as a full-time volunteer. When Kelsey learns that I'm no longer paid, she victoriously rejoins the March in Colorado Springs. I welcome her back, though without enthusiasm as I'm certain she'll soon find a new way to sow discord.

Colorado Springs is a success.[1] At our rally, Mary Barber, the director of Sustainable Fort Carson, shares how the US Army is responding to climate change. Congress may ignore 35 marchers demanding climate action, but if the US military speaks up, Congress might listen. Marchers are polite and attentive during Mary's speech, but one local heckles Mary, calling her a war monger.

Denver follows quickly on the heels of Colorado Springs. Chris Ververis, our state coordinator, has worked with a coalition of non-profit groups to pull together an excellent rally that's followed by a short march, then a meeting with Shaun McGrath, the regional head of the EPA. Later, Steve and I speak at a fundraising event that raises $500.

With Jimmy and I now unpaid, the March's fundraising goal is noticeably lighter. My weekly target is $7,500. Other marchers have stepped forward to help, too, launching a creative appeal called "Shave the Planet." As each new fundraising benchmark is met, another marcher's head is shorn. My folic-deprived dome is the first victim. By the end of the drive, nearly one out of every five marchers is bald. We raise around $15,000 and compost a whole lot of hair.

On June 18, we march north and east out of Denver on the first leg of our journey across the Great Plains. "Goodbye barren deserts and rugged mountains. Hello lush, rolling prairie!" I say to Steve as we suck down a second breakfast at a little cafe along the route.

"Well, for you maybe," replies Steve. "But I'm more than a bit nervous. Your lush prairie is infamous for scorching heat, high humidity, and killer tornadoes. Can't say I'm thrilled about any of that."

"Look, if we start real early, like 4:30 or 5:00, we beat the heat," I tell Steve. "And humidity? Great stuff. No more dry skin and chapped lips. Once you get used to it, humidity feels like a warm, moist blanket caressing your body. And tornadoes? There's nothing more thrilling

[1] http://gazette.com/marchers-stop-in-colorado-springs-to-tout-climate-change-message/article/1521250

than watching one of those bad boys tear through a field. If it comes at you, all you gotta do is dive in a ditch and it'll skip right over."

Steve looks at me skeptically, shaking his head. "You're so damn glad to be back on flat ground you've gone all stupid in the head."

We pause to eavesdrop on a conversation between two farmers at a nearby table. One mentions that he took a piece of equipment in for repairs "a couple-three weeks ago." I smile broadly at this distinctly Midwestern phrase, one I haven't heard in months. I say to Steve, "Am I glad to be back on the prairie? You betcha."

A strong storm hits as we arrive at our campsite at Prairie View High School in Henderson. I'm ready to pronounce a miserable end to our first day on the prairie when a brilliant double rainbow paints the eastern sky as the sun sinks into the Front Range. The hardship of setting up camp, cooking, and eating while dodging thunderbolts in a deluge is forgotten in this moment of sublime natural glory.

But that moment is itself short lived as a threat we've never considered strikes suddenly and forcefully. At the far end of the ball field, where my tent is pitched, the school's sprinkler system goes off. All I can do is watch and hope that the powerful stream hitting my tent every 30 seconds doesn't drench the inside. I'm grateful the fly is tight to the ground, but some of my gear gets wet anyway. I warn marchers that other sprinklers are likely to go off. They don't take me seriously until the next one pops up directly under Kim's tent, blasting a steady stream of water into the floor and soaking much of her and Liz Lafferty's gear.

Most marchers now hustle to move their tents. Some get drenched anyway. When the sprinklers are done beating us up, a second thunder storm rolls through. My welcome home to the prairie is complete, accompanied by a mere three hours sleep.

The next two days' marches are 17 miles each. Skies in shades of black, grey, and green bring more strong storms that include hail and funnel clouds. Our route follows the railroad tracks, and throughout the night the trains' whistles assault us every half hour or so. My ear plugs are worthless, the whistles tearing through them like a cougar shreds lamb flesh. The ongoing symphony of thunder and train whistles assures my third consecutive night of only a few hours of fitful sleep. I crawl out of my tent in the morning feeling utterly battered and exhausted.

To add to these joys, Kelsey wastes no time in launching her next campaign. She's upset that we use a low-mileage, fossil-fuel powered truck to move our gear. She lays out a plan for each of us to haul our own gear using strollers and rickshaws. She argues that if we want to drive home the urgency of climate change, all of us, not just her, have to make greater sacrifices. "We aren't suffering enough," she says as part of her sales pitch, to which Miriam replies, "I'm 71. I've marched every step of the way. I'm suffering all the time and I'm not pushing my gear in a baby carriage."

Kelsey's obsession with the gear truck is draining. Worse still, Jimmy rekindles his effort to gain control of the March. He argues that the website should be managed by "marchers" — and by that he means people who travel with the March, regardless of whether they often or ever actually march.

After we left California, I hired Ki to be our communications director. She's in charge of the website, which by any objective standard looks compelling and professional. Jimmy tries to convince me it's badly managed, yet I refute each of the points he raises.

Frustrated that I won't turn the website over to him, Jimmy and two other "marchers" drive — yes, *drive!* — to meet me on the route. I'm pissed off and feel ganged up on. I listen to them for a bit but my patience is thin. I interrupt and tell Jimmy he's trying to undermine a hard-working coworker, a trained professional who's doing the job we hired her to do. I tell the three of them that if they want to do something productive they should march.

They keep pushing. I lose my temper. "We've got Kelsey trying to get rid of the gear truck because we burn too much fossil fuel. We've got you driving all over the place, burning fossil fuel when you should be marching. I'm done with this conversation. Go back to your air-conditioned cafe if you want. I'm here to march," I say as I storm off.

The route from Roggin to Wiggins proves unworkable. There's no road, and we're stuck with two illegal options: the train tracks or the Interstate highway. After a few frustrating miles treading through the loose sand along the tracks, Steve opts to cut across a slice of range land and walk the Interstate. The rest of us continue along the tracks for about 15 miles. The challenge of walking through sand coupled with three days of sleep deprivation is devastating. I'm a walking zombie when I stagger into our camp on the lawn of a church. Sarah and John Jorgensen are there to meet me, and I can tell by the worried look on their faces that I've truly hit the wall.

Sarah comes to my rescue and finds a home stay. By mid-afternoon I'm sound asleep and remain blissfully unconscious for an impressive 15 hours.

Steve and I set out the next morning from Wiggins. "I don't think I've ever slept that long — ever!" I say. "Between storms, sprinklers, sand, and sleeplessness, I was like a dead man walking yesterday."

"Still lovin' your damn prairie?" quips Steve.

"Whatever," I snap. "It'll get better."

"I'm sure it will," says Steve. "But here's what's not going to get better: all this squabbling, these power struggles. Kelsey goin' off on the gear truck. Jimmy and Lala and these others who complain and make trouble instead of marching. I gotta tell ya, Ed, I've had enough. I know this isn't what you wanna hear, but I'm gonna leave the March."

I stop dead in my tracks. If there was one thing I never expected to hear, it was Steve saying he'd quit. He's been more vocal than anyone about the importance of marching every step of the way. And after hobbling along for 33 miles on crutches in the Mojave, I figured he'd drag himself on his belly to Washington, DC if he had to. I stand there, dumbfounded.

"I've been thinking, and I've got an idea," says Steve. "Marchers are talking about taking a four-day break to go to New York for the

People's Climate March in late September. Wouldn't it be powerful if one of us made that trek on foot, if the longest climate march in history met up with the largest? There's only one marcher who can do that, and that's me. It'd mean walking a lot of 25- and 30-mile days, and I'd have to haul my own gear. But I think I can do it."

I struggle for a response, and for lack of anything meaningful to add say, "Well, you hauling your own gear. That'd sure make Kelsey proud."

Steve laughs a bit, then gets serious again. "I hate the stupid shit that's going on in this camp, and I just can't take it anymore. I know it's even harder on you, Ed, and if you want out, you're welcome to join me."

We stand together on the side of the road, me squinting into the rising sun, leaning hard on my stick. I kick a rock in front of me and say, "I'd sure like to get away from this mess. But I can't. I can't leave what I started, no matter how bad it gets. I've gotta keep going, grind it out with everyone else, even the crazies and the power-hungry ones who want me to quit. Despite the few who are problems, there's mostly good folks here, and a whole lot of people back home who are counting on me, counting on *us*. I can't quit on them."

We walk on in silence. I'm sad. I don't want Steve to leave but I can hardly blame him. After a while, I say, "Well, I gotta admit, it's a great plan. I see the lead line in the story: 'In the battle for climate action, one battered warrior walks over 3,000 miles to join the largest climate march ever.'"

At Fort Morgan, the prairie skies continue their homecoming celebration with another huge storm, the biggest yet. We hurriedly clean up supper's mess as strong winds rip and tear at the trees above us. The wind is followed by intense rain that sends a torrent rushing through our camp — a stream of water over a foot deep. Some tents actually survive the deluge, marchers move others, and some are knocked over. Some marchers, including me, escape to a nearby hotel.

The next day, we walk by entire corn fields cut to the ground by last night's hail. Twelve hours after the storm, the hail still lies on the warm ground in thick piles, which from a distance look like snowdrifts. Locals I talk with tell me this is the worst string of storms they've ever seen, making the conversation about climate change a bit easier.

The clouds improve but the scenery doesn't. We pass feedlots that stretch along the road for over a mile, so far into the distance they fade from sight. Marchers — omnivore, vegetarian, and vegan alike — are appalled by this wasteland crammed tight with the living corpses of millions of cattle.

In camp that evening, a few marchers become vegans in response. One young convert poetically declares, "Meat is like the flesh of babies and dairy the lubricant that grinds the very wheels of hell." I appreciate the colorful metaphors but argue that the problem's not meat but the way it's raised. Yet reason is lost on urban innocents, who've only known meat as a pretty package on a grocery store shelf, who now stand in shock as their eyes are pried open to one of industrial agriculture's many dark secrets.

The next day, Steve sets out on his own. We shake hands and I wish him luck. I tell him I'll see him on the coast. I know he'll make it, but I also know it'll be rough, a dangerous haul alone on highways that will get busier the farther east he goes. I reassure him that the harshness and emptiness of the high Colorado plains won't be an issue in Iowa. "The tall-grass prairie is a friendlier landscape, Steve," I say. "You'll see. You'll be there soon enough, at least a week before the rest of us. Iowans will treat you right, even though they'll think you're a weirdo, pushing a baby stroller along the highway, and that funny accent of yours. And Iowa ain't flat, so get that notion out of your head."

I watch Steve set out on his 1,500-mile solo walk, watch him grow faint in the distance, pushing that stroller, looking every bit as out of place as the early settlers must have looked in their covered wagons. Maybe someday, we'll learn how to live with the prairie, not just come in, tear the whole thing up, and let it wash away to the Gulf of Mexico. I think of Loren Lown, a friend and conservationist from Des Moines, who was quoted a few years back saying, "We have left less than one-tenth of one-percent of our prairie. The rest of it died to make Iowa safe for soybeans."[2]

I've been looking forward to Nebraska because of the prominence of the Keystone pipeline in the fight for climate action. Our state coordinator, Anna Wishart, has done amazing work lining up local contacts and assembling an impressive amount of donated food, much of it locally grown. As we enter Nebraska on June 30, I'm excited about our prospects for making a difference here. But the two marchers I'd grown closest to — Shira and Steve — are gone. Conflicts with other marchers threaten to gain momentum, and I feel increasingly isolated and uneasy as we approach the halfway point of the March.

---

[2] Loren Lown (p. 139 in Richard Manning 1995, Grassland: The History, Biology, Politics, and Promise of the American Prairie, Penguin, New York, NY)

CHAPTER 24
INDEPENDENCE

"There never yet was, and never will be, a nation permanently great, consisting, for the greater part, of wretched and miserable families."

— WILLIAM COBBETT

With Steve gone, my marching day becomes increasingly contemplative. I often avoid the route Sarah and Anna have laid out and tack on extra miles to walk quiet roads or detour through small towns in search of a cafe with Internet and large servings of bad food.

Western Nebraska is sparsely inhabited but under-appreciated, exuding a wealth of sights, sounds, and smells too abundant to catalogue. The lack of traffic liberates me from fear that the next car or truck barreling past could maim or kill me. My focus shifts from survival to the alluring world around me — and to the even more alluring world within.

Marching becomes meditation, my footsteps the mantra. I see the fields, ditches, trees, irrigation pivots, fence lines, homes, and out-buildings. I hear the dog bark, the cow bellow for her calf, the cardinal sing to his mate, the warm breeze rustle the chest-high corn. I smell the white clover, the fresh-cut hay, the comfortable scent of horses, the acrid pungency of too many hogs. All this and so much more drifts through my senses in slow motion — visual, audial, and olfactory b-roll, the canvas for the actual performance of life itself.

My mind focuses on the repetitive, rhythmic crunch of shoes on gravel. It clears my head and brings a sense of peace. I recall the meditation course I took at age 16. The ten-minute introductory session induced an unexpected inner calm that remained with me the rest of the day. Nothing bothered me — not the blackberry thorns that tore at my skin as I harvested the plant's fruit, not my mother's nagging, not my brother, Bill, calling me names for sport.

Years later, after a long day hitchhiking through the French Alps, I landed high up in the mountains at a Buddhist monastery. Sitting for hours with the monks as they chanted "om," the sound playing off ancient stone walls that once housed Catholic monks, I noticed how the mantra would roll through six or more unique tones in one recitation.

Decades later, on a work day during my campaign for governor, I thought about that experience as I made tiramisu at Cafe Dodici, an Italian restaurant in Washington, Iowa. The restaurant's young, artistic chef showed me how to blend the egg yolks. "Watch how many different shades of yellow they go through, like 15 or 20," he explained excitedly. "It's awesome, as if you're watching the universe unfold in a mixing bowl. But you'll need some tunes to really bring it home," he said as he flipped a switch, sending loud rap music blaring through the kitchen.

Om. Egg yolks. Footsteps. There are endless aids to center oneself on the path to enlightenment. But a mantra isn't stagnant white noise. It's alive, rich with motion and texture. My right heel's first contact with gravel produces a deep tone. There's a sudden decrescendo as the foot begins to roll forward. The pitch and volume rise as my weight shifts to the ball of my foot and the left heel moves into place to repeat the pattern. Every four steps, my walking stick punctuates the rhythm with a sharp sforzando as it grinds into the loose gravel. Right. Left. Right. Left. Right. Left. Seven million times — and that's only the beginning.

Like waves breaking on a beach, my footsteps roll through gravel, through Nebraska, toward infinity, toward eternity. At times like this, my mind seems to get it. The technique and purpose of meditation — directing the hungry soul toward the peace that comes with knowing one's higher self — is so simple and so transparently important. Yet more often than not, my mind remains restless, distracted by both beauty and ugliness, unable to focus on the deeper truth that transports one beyond appearance, beyond pleasure and pain.

Forty years ago, my first meditation was a uniquely powerful experience. But life's pressing demands lured me away from the pursuit of inner peace. Perhaps had I continued to meditate, continued to cultivate the balance that such practice brings, I'd be able to manage the March's turmoil with more dignity. Perhaps meditating during my solo walks on backroads might yet help me deal with the challenges ahead, help me avoid tempestuous outbursts.

A dog barks. I re-enter the world of the senses. What kind of dog is that? Is it on a leash? Does it bite?

A bird sings — wren or finch?

Will the cafe in the next town serve real butter?

I hope I don't run out of wet wipes today.

I suck at meditation, even under the tutelage of a guru as patient as western Nebraska's gravel roads.

"Surprised how quickly I've adjusted to walking alone," I tell Steve in a text message. He and I keep in touch most days. On July 4, I call and wish him a happy Independence Day. "So, traveler. How ya feelin' about your independence now?" I prod.

"Well, I won't lie to you, Ed. I've been walking extra-long days — thirty miles or more — just to put some distance between me and you marchers," Steve laughs. "Besides, I worry that Creekwater will try to catch up with me, walk with me, even offer to cook dinner."

In a community of quirky types, Mark Creekwater is perhaps the quirkiest. Mark's surname of choice reflects his almost religious belief that, despite humanity's desecration of water, one has a right, even an obligation, to drink fearlessly from any stream. Mark's a strong walker but his views on water become problematic when he helps in the kitchen. Occasionally, he tries to use creek water to wash dishes, or worse, to brew some concoction he'll attempt to entice marchers to eat or drink. No one gets sick, but monitoring Mark's evangelical aquatic mischief requires constant vigilance.

"Ha!" I say. "Yeah, he's the one person who could probably keep up with you. But you'd be dead of some water-borne ailment before you hit Illinois."

"Right. So do me a favor and keep Creekwater with you," says Steve. "I've got my hands full dealing with long days and these prairie thunderheads you love. But hey, I've got an idea on how to handle the next storm — I'll just flip my cart over, crawl under, and ride it out. What do you think?"

"You're nuts!" I warn. "You can't seriously think getting fetal under a cart is gonna keep you and your gear dry. And that cart's made of metal. Did ya think about that? There's a lightning bolt out there hungry to char-broil your dumb ass cowering under a metal cart."

"Yeah, that probably wouldn't work," Steve mumbles. "Ya know, even though I'm glad I don't have to deal with camp, it sure was nice to finish a day's march knowing Sarah would have everything set up. Maybe after more time walking alone I'll appreciate what a group can accomplish. Maybe I'll get to the point where I'll feel the loss of community. Right now though, I'm glad to have some peace and calm."

Culbertson, a town of 600, swells to several thousand people for its annual Independence Day celebration. We're thrilled to have been invited to join the parade. Some of us play instruments and sing while others carry signs and a banner. Folks along the parade route seem genuinely receptive, or at least politely amused, and we connect with our biggest audience since Los Angeles.

After the parade, I immerse myself in Culbertson's holiday fun. I watch the horseshoe competition for a bit, then stumble on a luncheon to raise funds for the new library. I drop $10 for a modest meal and grab a seat at one of the tables, striking up a conversation with a woman who introduces herself as Corky Krizek. She and her family live in McCook and they saw us in the parade. Corky's got the usual questions about shoes and weather, then asks "Have you lost much weight?"

"Yeah, dropping 24 pounds in the first two months was one of the biggest surprises," I tell her. "I rip through calories like a twister through a cornfield, and I'm craving meat like there's only one pig and one cow left on the entire planet."

"Well, when you get to McCook tomorrow, you'll be only a few blocks from our place. Stop by and I'll make you a big steak dinner."

I thank Corky, then wander around for another couple hours, reveling in the nostalgia of all that's good and wholesome about America: family, food, fun, and a robust love of land and country. I think about the myriad ways in which everyone here, each of these several thousand people, need each other, how their lives are woven together in so many essential ways. July 4 is not so much a celebration of America's independence as it is Americans' interdependence.

Perhaps that little girl in the red dress over there, the one darting around the playground with her friends, will only overcome her learning disability with the after-school reading program at the new library.

Perhaps that old farmer I saw tossing ringer after ringer at the horse-shoe contest had an accident last fall and was only able to get his crops in with the help of his neighbors.

Perhaps the Climate March wouldn't have even made it out of California without the kindness and support of hundreds of people. Yeah, I'm certain of that.

Sure, Americans should celebrate winning our independence from England, even though things probably would have turned out about the same whether we'd fought a war or followed the more diplomatic path of our Canadian neighbors. Sure, we should celebrate the fact that, over the course of 238 years, no foreign power has come close to invading our country and subjugating our people.

But meanwhile, we've bought the notion that independence means being able to do whatever we want, whenever we want, without anybody's help. The percentage of Americans who now live by themselves has swelled from five percent in the 1920s to 27 percent today.[1] That's not independence. That's isolation. That's the face of loneliness — and though it may be hard to measure, those who study these sorts of things claim loneliness has increased dramatically over the past twenty years.[2]

---

[1] https://www.washingtonpost.com/politics/more-americans-living-alone-census-says/2014/09/28/67e1d02e-473a-11e4-b72e-d60a9229cc10_story.html?utm_term=.55c42dd34fb5

[2] http://fortune.com/2016/06/22/loneliness-is-a-modern-day-epidemic/

On July 4, we celebrate our independence from foreign powers. The rest of the year, we celebrate our independence from each other. Meanwhile, we've failed to notice that America has succumbed to a gradual invasion, a more insidious subjugation. Through the clever manipulation of laws by greedy men (yeah, again, they're mostly men) and our own complacency, national chains and big corporations now dominate our economy. It's increasingly difficult, almost impossible in some professions, for hard-working men and women to harness their talent, energy, and passion to realize the American dream and earn a living as a farmer, business owner, or entrepreneur not beholden to some distant corporate overlord.

While we cheer the parade vehicles made in Japan, wave our tiny flags made in China, and catch little pieces of candy made in Mexico, the wealthy and powerful quietly consolidate their control over our lives. They do this in large part through buying off America's political leadership, both Democratic and Republican, and solidifying their control of our lives through manipulative advertising. We fail to notice that this unholy alliance of corporate and governmental power has eviscerated anti-trust laws, gutted protections against the formation of monopolies, allowed foreign corporations to buy our farmland, and enacted trade treaties that ship our jobs and factories overseas. When the powerful interests that benefit from these laws run aground because of their own greed and stupidity, our politicians simply provide taxpayer-financed bailouts to banks, car manufacturers, and other industry giants deemed "too big to fail."

The way out of this loss of independence is through recognizing, celebrating, and building upon our essential interdependence. Buying our food from farmers we know and trust. Supporting businesses owned by people who live and work in our town. Using cash instead of credit cards, since the small business owner in the middle gets dinged badly by the credit card company. Doing more with barter.

The long road that led us from America's former independence to our current dependence — and the difficult path out of dependence through interdependence — is our only hope if we are to win both the race against climate change and the struggle to regain our democracy.

The march from Culbertson to McCook is a tough 21 miles. The mercury hits 101° as I land at Corky's home in mid-afternoon. She, her husband, daughter, and a friend are hunkering down out of the blistering heat. It's too early for a steak dinner, but I've arrived just in time for a big slice of Key lime pie. Corky offers me a second piece and I don't refuse.

"This is just what my body's been craving! I'm ready to walk another 21 miles," I jest. "Who's in?"

Corky and her friend laugh, then our conversation turns serious. She and her family are evangelical Christians and solid Republican voters. "I've been in Nebraska for almost a week and I've only met one Democrat," I laugh. "But most of the Republicans I talk with think we're doing a good thing with this March, think it's important that we're trying to raise awareness about climate change, even though most Republican elected officials deny climate change is even happening."

Corky shares my concern that our planet is in peril. "Something needs to be done. I don't know exactly what's causing climate change, but I know that we can't go on the way we are, destroying our Earth with so many chemicals and waste products."

"What about your faith?" I ask. "Does that speak to the problem?"

"More than anything, Jesus was concerned about the welfare of people, especially the poor and the needy," Corky tells me. "He dealt mainly with their spiritual and physical needs, but I know Jesus wouldn't want us abusing the land like we're doing."

Corky was serious about the steak dinner. She also offers to put me up at the Chief Motel in downtown McCook. "That's sure kind of you," I say. "I'm truly grateful. It'll give me a chance to get caught up on some work, too. But I have to swing by our camp first and pick up a change of clothes."

"It's a good mile or more away, so I'll drive you there, then drop you at the hotel," says Corky. She and I hop into her Honda Odyssey and head to camp. I figure we'll get out and walk around when we arrive, but Corky prefers to stay in the car. We take a slow drive through camp as I give her a guided tour of the various points of interest — solar collector, solar ovens, Eco Commodes, kitchen truck, gear truck. It's an odd feeling, sitting in the passenger seat of an air-conditioned van, windows closed, driving through the camp that's normally my home as marchers peer back and wonder what's going on. Whatever critical assumptions they might be making, I doubt even Jimmy, Lala, or Kelsey would have guessed that I was on my way to a steak dinner at a hotel.

I thank Corky as she drops me at the Chief. I've slept at a dozen or so hotels and motels during the March, often on the floor, usually packed in with three or four other marchers. To have a room and bed to myself with air conditioning and a decent Internet connection is a royal treat. I work for a couple hours until Corky shows up with steak, potatoes, vegetables, bread, and real butter. It's such a huge meal even I can't finish it. I pack half the steak into my satchel for tomorrow's lunch, and in an act of incomparable cruelty send a photo of the meal to Steve. "See what you're missing by not sticking with the March?" I taunt.

I fall asleep shortly after 8:00 and enjoy a quiet night undisturbed by traffic and trains. Fuller's Cafe just down the road opens early, and after my usual massive breakfast, I'm on the road by 6:20. The March campsite at Karrer Park is two miles east of the hotel, and most marchers are still asleep when I arrive. Curiously, several are sprawled out on the sidewalk, including Jimmy, who's curled up with only a blanket. Sarah is awake but bleary-eyed and upset to the point of tears. She'd been clear with park officials about disabling the sprinklers, but nonetheless they went off at 2:00 a.m. On such a hot, sticky night with no rain forecast, tents were set up without flies. When the sprinklers hit, marchers were defenseless, some soaked before they knew what was happening.

I feel bad for Sarah and horrible that marchers endured a second sprinkler incident in less than three weeks, especially while I enjoyed a comparably luxurious night. I've shared in most of the March's misery, and occasionally more misery than others. But the pleasure of last night's room, shower, and steak dinner are now lost in a sea of guilt.

I march with others this morning and learn of a hastily organized vigil yesterday at the train station in McCook. A year ago, forty-seven people were killed in Lac-Mégantic, Quebec when an oil train derailed and exploded. Some marchers made a quick decision to memorialize the tragedy, sitting quietly in the station's lobby while reading the names of those killed, ringing a bell for each victim. It was a powerful action and garnered a favorable story in the *McCook Gazette*, complete with photos.[3]

Our time in McCook is emblematic of the conflict central to my role in the March. I walk every step of every day while struggling with an intense workload as director, fundraiser, and talk show host, and there's no way I can accomplish it all on minimal sleep and inadequate calories. On top of that, it's important that I take time to talk with people I meet, especially those not part of the choir. So I justify staying at a hotel and eating steak while other marchers camp by a noisy highway on a hot summer's night, sleepless and drenched by sprinklers. Today, that justification feels hollow.

---

[3] http://www.mccookgazette.com/story/2098399.html

CHAPTER 25
CROSSING KEYSTONE

"Loneliness, thy other name, thy one true synonym, is prairie."

—WILLIAM A. QUAYLE, **THE PRAIRIE AND THE SEA**

It's hard to understand how Kelsey can say with a straight face that we aren't suffering enough. I may have escaped crucifixion by sprinkler last night, but I make up for it tonight during an intimate encounter with mosquito spray.

Bartley's town square is our campsite, and most of us are bedded down by dusk. I'm drifting off to sleep when the sound of a spray truck jars me into consciousness. Through the tent door I see the toxic cloud of joy heading our way. I rush out wearing only shorts, waving my arms wildly. The driver stops, turns off the sprayer and says, "I figured you could use a little extra dose of protection against these skeeters."

"That's really kind of you," I reply groggily. "It's just that some of our marchers are allergic to the stuff." I realize I'm lying. But I'm half asleep and convince myself that a meticulous commitment to honesty is demanded only of the fully conscious. "Besides, that spray'll just drift right through these tents, make us and our gear all sticky," I add truthfully.

"Not a problem," he says. "Just didn't want you to get eaten alive. Hope you get some sleep. I sure do admire what you're doing."

I thank him again as I crawl back into my tent. "Maybe having spared most marchers from being poisoned will earn me a few brownie points," I mumble as I contemplate the prospect of a decent night's sleep, the fragrance of malathion lingering in the hot, summer night's air.

Like the rapid succession of plagues sent to torment Biblical Egypt, next day brings a new form of suffering. Both our route and campsite are so infested with ticks that I pull 14 off me by the end of the day. Jane runs a piece of white paper over the weeds alongside the road, flips it over and counts nearly 50 of the blood-sucking monsters!

None of us have ever seen such a concentration of a creature that, from an anthropocentric point of view, is a prime contender for *World's Most Evil Insect*. I joke that perhaps they've come here from across the country for some kind of national tick convention. On a serious note, I warn marchers to conduct a thorough tick check before crawling into their tents. I share the story of a young man on the Peace March who was hospitalized in Des Moines after a tick embedded itself in his ear.

Ironically, as we walk to raise awareness of the risks of climate change, we elevate our own exposure to these risks, including the growing frequency of tick-borne diseases.[1] Just six weeks ago a few hundred miles from our route, Johnny Mitzner of Delaware County, Oklahoma died of Heartland virus, compliments of a tick bite.[2] Living almost entirely outdoors, marchers have reason to be concerned. Fortunately, none of us are infected with any of the half-dozen tick-transmitted diseases on the rise in recent years.

A few days later, we barely avoid another opportunity to suffer. We missed a monster hail storm by 24 hours that stretched eight by 100 miles. I'm enjoying my lunch break on a stump next to a cornfield, admiring the tall, healthy crop while eating two peanut butter and jelly sandwiches. As I walk east, I notice torn corn leaves, then broken stalks. By the time I reach the end of the field the crop has disappeared almost entirely — eight-foot tall corn reduced to stubble, cut nearly to the ground. I've seen hail damage before, but nothing like this.

We arrive at our campsite in a park in Gibbon to an almost warlike scene. Large hail stones have broken glass and smashed cars.

---

[1] https://www.scientificamerican.com/article/tick-borne-diseases-on-the-rise-thanks-to-global-warming/

[2] http://www.newson6.com/story/25655554/son-of-man-who-died-of-tick-borne-virus-warns-of-dangers

Siding has been ripped off buildings, paint stripped from houses. In the park, mature trees that normally would provide shade are half naked, their foliage shredded by hail. I try to imagine what would have happened to our tents, vehicles, and bodies had we camped here the previous night when the storm struck. Maybe we finally would have suffered enough to satisfy Kelsey. Quite possibly, between the absolute destruction of our tents, severe damage to our vehicles, and who knows how much carnage to tender marcher flesh, it would have marked the dramatic and painful, perhaps deadly, end of the Great March for Climate Action.

One Earth Village has lived with an uncomfortable peace since Ki and I prevented Jimmy from taking control of the website. For the first time in weeks, a handful of marchers call for a camp-wide meeting. I'm resentful and have no interest in an unsanctioned and probably unnecessary meeting. I'm worn down, my stomach hurts, and there's still work to be done. I just want to rest after another difficult 15-mile day.

I don't know what to expect. I'm only a little surprised when Jimmy, Kim, Liz, and Lala again accuse me of mismanaging money. These are four "marchers" who almost never march. Even Lala, once one of our strongest marchers, now spends most of her time doing what, I'm not sure — apparently plotting to make my life miserable.

The four want greater control over the March's finances and demand a plan for cutting expenses. I sit there and say very little, feeling strangely detached from the conversation. An odd boredom washes over me as one after the other take shots at a mostly deflated Ed punching bag. I want to say, "With all the energy you guys have after sitting around all day, surely you can come up with a more challenging target than me." But I keep my mouth shut.

Then something unexpected happens. Izzy speaks up. He defends me, says I'm working hard, and as long as the March has enough money to keep going, he's got no worries. He's tired of the drama and in-fighting. John Jorgensen speaks up next, then David Zahrt. Both reiterate Izzy's perspective. I know plenty of marchers don't like the way I've been treated and are sick of the attacks and complaining. But until now they haven't spoken up. The meeting concludes with a "decision" that means nothing, since it's how Shari and I have been managing finances all along.

The next day brings an 18-mile march. My stomach is a tight ball of intestinal badness that only gets worse as the miles roll along. On two occasions, my bowels command me in no uncertain terms to dive into a ditch and promptly evacuate the remnants of last night's couscous and kale.

I walk slowly, unsure if I'll be able to push on through this intense discomfort. Not trusting tonight's supper to settle well with my gut, I stop at a restaurant a mile or so from camp where I order a bowl of mashed potatoes. The young waitress is fascinated by the March and sympathetic with my condition. She insists on paying for my modest meal. Acts of generosity have become a daily experience in Nebraska.

When I arrive at camp, I'm a visible wreck. Marie insists on setting up my tent and brings me peppermint tea. Given the renewed hostility I felt in camp last night, her kindness is particularly appreciated. I slump into my tent where I spend the next 12 hours recovering.

Kindness. Acceptance. Encouragement. I encounter so much of these in Nebraska — some from marchers, mostly from strangers. There's the pickup truck driver who pulls over to ask what we're doing. When I explain he says, "I'm a conservative Republican, but I'm also a conservationist, and we absolutely have to do more to protect the environment."

There's the man driving a minivan sporting a pro-life bumper sticker who rolls down his window and asks, "Are you with those people protesting the pipeline?"

I can't tell whether he's for or against the pipeline, but respond, "Yeah, that's us. We don't like that pipeline one bit."

"Well, thank you for what you're doing," the man says with sincerity. "I hope we can stop it. It just ain't right."

Across western and central Nebraska, nearly all our encounters with local people are favorable. The majority are conservative Republicans. While many people are initially suspicious of this odd group of ragamuffins walking through their town, curiosity about why we walk opens the door to conversations and, in many cases, to the discovery of common ground.

July 19 is a much-anticipated moment — the day we cross the proposed route of the Keystone pipeline. We have a short trek today. A couple dozen local people join us for the ten-mile walk to the Energy Barn — a small solar- and wind-powered structure built defiantly in the path of Keystone. Marchers are invited to be first in line at the lunch table. They descend on the offerings like a swarm of hungry locusts, and I cringe at the possibility that they might not leave enough food for others.

The afternoon is festive and full of activities. Billboards are painted for landowners who live along the route. A flag reading "No Permit, No Pipeline" is erected next to the barn. We hear a handful of speeches. There's more food, and as dusk arrives the movie *Above All Else* plays in the Energy Barn. I skip the movie and opt for sleep, retiring to our camp at a nearby farm.

As Nebraskans share their concerns about Keystone's threat to their land and water, particularly how an oil spill would impact the Ogallala Aquifer, I feel fortunate that Iowa is not in the crosshairs of a pipeline project. Ironically, we learn today that a Texas company has announced plans to build the Dakota Access Pipeline diagonally across Iowa. The company is Energy Transfer Partners and its owner, Kelcy Warren, is one of the wealthiest men in America. Warren's plan is to bury the 30-inch pipe through 350 miles of Iowa farmland, transporting up to 24 million gallons of oil a day — enough to create carbon emissions equal to that of 30 new coal-fired power plants.[3] Not only would this pipeline threaten Iowa's waterways and exacerbate the global climate crisis, it would use eminent domain to forcibly take land from farmers and potentially damage crop yields and drainage systems.

It would pass less than 20 miles from where I live in Des Moines.

I feel as if my home itself is under attack. I want to quit the March and help organize efforts to stop the pipeline. But I'm committed to finishing what I started, committed to grinding through the remaining 16 weeks and 1,200 miles to Washington, DC.

---

[3] http://priceofoil.org/2016/09/12/the-dakota-access-pipeline-will-lock-in-the-emissions-of-30-coal-plants/

Nebraska may be relatively flat, but the past three weeks have been a physical and emotional roller coaster. Especially now, with the added worry of a possible oil pipeline across Iowa and the excitement of seeing Grace, my head and heart undulate in sync with the rolling prairie.

Thoughts of Grace keep me going. I miss her so much. We talk on the phone every few days. Knowing that in a few weeks we'll be able to spend a week together in Iowa encourages me onward.

I think of how the prairie reflects my own emotions and desires. Such an empty, lonely place — but one that encourages thoughtfulness, introspection, and patience. In just over three months, this 246-day odyssey will have accomplished its mission. As I walk over each new rise on the distant prairie horizon and my eyes look ahead to the next — one mile, or three, or perhaps five miles away — I think, "how many more such rises before I see my Grace smiling at me, welcoming me home for a short visit before the final thousand miles of this journey?"

A MEMOIR FROM THE GREAT MARCH FOR CLIMATE ACTION

CHAPTER 26

ACROSS THE WATER

"Love is blind."

— GEOFFREY CHAUCER, "THE MERCHANT'S TALE"
IN CANTERBURY TALES

In the spring of my 22nd year, I flew to Ireland with $800 in my pocket and fifteen pounds of gear on my back. For the next eight months I traveled through Europe and the Mideast living mostly on air, kindness, and creativity. It was a journey with no specific destination or goal, an unstructured spiritual quest seasoned with wanderlust. I often worked in exchange for room and board, staying at monasteries and other venues attractive to the spiritually curious — places like Taize, L'Arche, an Israeli kibbutz, and the compound of a cult disguised as a latter-day hippy commune.

I also stayed with "friends," defined as anyone I'd spent ten minutes with who'd put me up for the night. Sometimes, I'd make my bed on a park bench, in the corner of a train station, behind an organ in an airport chapel or, more comfortably, on a mattress on the sand with Bedouins under the rich, starry-black night of the Sinai Desert.

The adventure began with a two-week walk across southern England along the Pilgrims Way. I followed the route of the colorful cast of Chaucer's *Canterbury Tales* through amber fields of rapeseed and forest floors carpeted with bluebells. I camped in places I imagined the Miller, the Pardoner, the Friar, and Chaucer's other heroes and ruffians had laid their heads after a day of modest walking and heavy drinking.

My tent was a rain poncho staked to the ground, a stick on either end raising the middle just enough to direct rainwater off the sides. When it rained — and it rained often — the poncho kept either my head or feet dry, but not both. I was certain Chaucer's pilgrims had enjoyed more comfortable lodgings.

A few weeks later while hitchhiking through Belgium, I thumbed a ride with a pleasant woman my age. We chatted about God and life, and by the end of the thirty-minute ride we were "friends." She invited me to visit. A week later, I did. Two days later, she fell in love with me.

Her name was Annika Duthoo, a born-again Christian who unabashedly declared both her commitment to pre-marital abstinence and her desire to eventually bear seventeen children. I liked Annika but wasn't in love with her. I enjoyed her company but had no desire for marriage, a horde of screaming fidgets, or her brand of Christianity. What I wanted was a traveling companion, and when Annika suggested we spend her six-week vacation in Scandinavia, I agreed.

The adventure didn't last long — less than an hour, in fact. Annika's mother and father objected vehemently to our trip, and when the emotional impact of parental scorn caught up with her twenty miles out, Annika broke down. She sobbed uncontrollably as we drove to the home of an evangelical minister to seek his consolation and guidance. They met privately while I wandered off to sit under a bridge and ponder my predicament.

Filing my nails on a slab of concrete to the rhythmic pulse of overhead traffic, I knew there was one obvious, intelligent thing to do: grab my backpack, hitch a ride eastward, and let this mercifully brief spate of relational drama fade into the setting Flanders sun.

But I couldn't leave without at least saying goodbye. Annika emerged from the minister's house, and before I could announce my intention to leave, she presented a new proposal. We'd stay at her parents' home for the next six weeks and make day trips to various points of Flemish interest. I glanced at my shoddily filed nails, stared at the open road beckoning me toward the heart of Europe, and in what would prove to be the first of many bad decisions I'd make in the realm of love, I accepted Annika's offer.

Our time together was a blast until I wrecked my back in a game of soccer. Annika's parents let me hole up for another six weeks to recover, but with little improvement. The family doctor recommended I go somewhere warm. Ignoring the fact that, from a medical perspective, this advice was absurd, I took a train south and stayed with various friends and strangers along the Mediterranean Sea for two months.

With still no improvement, Annika offered to loan me the money for a plane ticket home to the States. I spent my last cash on a flight to Brussels and landed there on a cold, wet night in November, my back screaming epithets at me. I called Annika from the airport. She told me she'd changed her mind, didn't want to help, and didn't want me to visit.

I was devastated, physically a wreck, alone in a foreign country — 3,500 miles and one very large ocean from home. I had no money and was in near constant pain. I could sit for only short spurts before my back sent intense nerve pain pulsating down my right leg.

My singular obsession was to return home and get medical attention. In a strange twist of circumstances, I befriended a Russian dissident who helped me secure passage across the English Channel. I then found my way to London where my friend Phil Sowden lived as a squatter in an abandoned apartment. Each morning, I'd catch a train from Phil's place to the London docks, determined to stow away on a ship crossing the Atlantic.

News of my dockside desperation came to the attention of Maureen Impey, an Irish cousin living in London. Maureen tracked me down and chastised me severely. She then bought me a plane ticket and sent me home, mercifully ending a miserable adventure that had started with such promise.

The Climate March is now less than 30 miles from Lincoln. We score a peaceful campsite in a park far from the auditory clutter of trains and traffic. I pitch my tent away from others, down a gentle slope by a muddy, ambling creek. The night is as black as prairie loam, the air resonant with the music of insects and frogs.

There's a soft splash — some fish, reptile, or amphibian — which class of creature, I can't tell. It's followed by a series of deeper, softer splashes, as if the creek's heart is pumping life from within the rich mud to its surface.

Tonight's white noise is comforting. Listening intently to the pulse of the water, resisting the temptation to sleep, I think of the human heart and reflect back 34 years to that European adventure turned sour. Like the March, it started with a walk. Like the March, it lasted eight months. Unlike the March — I hope — it ended in physical agony for me.

What I've come to realize is that it also ended in an agonizing heartache for Annika. Back then I had pity only for myself, feeling scorn and contempt for Annika. I was outraged that she would lure me back to Belgium, penniless, then abandon me. Now, across the span of many years and thousands of miles, I feel pity for her, too.

Why is falling in love so often a one-way street? Over the years, I've traversed both sides of that path. Now it's easier for me to understand Annika's uncontrollable sobbing, her indecision, her irrational compulsion to cut me off suddenly and emphatically.

Of all the forces that drive human activity, love is both the most powerful and least comprehensible. What psychic dysfunction compels the Universe to make a person feel deeply attracted to someone who doesn't feel the same? It's as if God finds some sadistic entertainment in setting up both parties for failure — one convulsed in heartache, the other wallowing in guilt.

"Damn you, Universe, or whoever you are," I say aloud to the insects, frogs and whatever being stirs below in the creek. "You've made humanity as dependent upon love as it is upon air, water, and food. Damn you. I'm tired of having my heart broken, tired of breaking someone else's heart. So there."

I slip back into silence, content that I've bitch-slapped the Universe's highest existential power, feeling smug that I sent it cowering into some remote hole in a distant galaxy. At least from the perspective of a privileged American, the quest for love is tougher even than the daily quest for food and water. Many times I've been tempted to give up, to stop trying, to resign myself to a life without a love-partner.

But just as we persevere in the fight against injustice, we persevere in the search for love. We have no choice. Life without both struggle and love is the rhythmless dance of the living dead. It is existence without heart, water without movement, night without desert stars or prairie fireflies. So, for the sake of love, for the sake of justice, for the sake of life itself, we keep going, one day at a time, one step at a time, hoping to get it right — if not this time, maybe the next, or the next, or the next.

I drift off to sleep feeling grateful and blessed that, with Grace, I found a love that is mutual. Together, Grace and I finally got it right.

The March is now less than 200 miles from Des Moines, and several friends join me for a day or two. Charles drives past me on the road and doesn't even recognize me. At camp, he says, "You look pathetic. I can't believe how skinny you are. Here, eat these sardines," as he hands me three cans of protein-rich medicine.

"Doctor's orders?" I enquire.

"Yeah, and cheaper than anything I'd prescribe from Big Pharma."

"Ok, doc, so here's what I really wanna know. The bottoms of my feet have been aching for the past hundred miles or so and getting worse. I thought it might be because they grew a whole size and my shoes were too tight. But Karin came to visit a few days ago and picked up a new pair for me. They're a size larger. I thought that would do the trick, but it hasn't helped."

Charles looks at my feet, pushing here, prodding there. "Does this hurt? Does that hurt? How about this?" I answer yes to each question.

"Plantar fasciitis," he announces. "The only way you're going to cure it is with rest. You've got to stay off your feet for a couple weeks."

We look at each other for a few seconds, then both start laughing. "Well, I took your advice about eating meat, but I can't take your advice this time," I say.

"Yeah, I know. I guess you just end the day with epsom salt, massage, and ibuprofen," says Charles. "But if it keeps getting worse, which is likely, you may be forced to quit whether you want to or not. I know you don't want to hear that, but you're not Superman. And you're no spring chicken."

I thank him for his candor and the sardines. Charles sets up his tent next to mine. I'm hungry for news from back home so we talk until well past my usual bedtime. I'm grateful for the friends who've supported me and our mission, though Charles has now planted a seed of doubt about whether I'll be able to continue marching.

Our rallies in Lincoln and Omaha go well, but for me they feel like distractions, bumps in the road, obstacles standing between me and Iowa, between me and Grace. Over each horizon, I strain my eyes to try and pick out the path of the Missouri River in the distance.

The river doesn't show itself until we walk into downtown Omaha. Glistening off her brownish waters, I imagine the sparkle in my Grace's brown eyes as she greets me in about a week, perhaps in Greenfield, just west of Des Moines.

Halfway across the river on the Bob Kerrey Bridge, we stop to chant "California, Arizona, New Mexico, Colorado, Nebraska, IOWA!" The moment brings tears to my eyes. I realize more fully than ever how important is the place one calls home. We finish the crossing and descend into the Missouri River's broad floodplain, the fertile edge of this verdant crescent nestled between America's two mightiest rivers. I know I'll only be in Iowa for three weeks, and only one of those with Grace. But for now, it's enough.

CHAPTER 27

"DON'T BE STUPID"

> "To get politicians to listen, you often have to do something demonstrable. I lived in a chicken coop. You're marching across the country. I'm not sure which is more out of the box."
>
> — DAVID OSTERBERG

Rising dramatically out of the broad Missouri River flood plain are Iowa's Loess Hills, formed 18,000 years ago by wind-blown soil as the glaciers retreated. Beyond the Hills lies the fertile, green patchwork of western Iowa's rolling prairie. Many marchers assume Iowa is flat. Today's 17-mile trek from Council Bluffs to Treynor convinces them otherwise.

"You're not a bad walker for a guy who sits on his duff all day," I tease my friend, David Osterberg, as we lean into another climb.

David joined the March last night and plans to walk to Des Moines. He directs a think-tank called the Iowa Policy Project. Earlier in his career, David taught economics at Cornell College before serving in the Iowa House, where he helped author several pieces of landmark environmental legislation. Most notably, David gained national recognition for living in a converted chicken coop for five years.

After returning from the Peace Corps, he felt driven to make a statement about how wasteful our society had become. His henhouse was tiny by modern American standards, but comparable in size to what millions of people around the world call home. Predictably, local government officials tried to evict David. He sued and the district judge ruled in his favor, elevating David to hero status among libertarian-types across the country and launching his successful campaign for the Iowa Statehouse.

One can argue the wisdom of making a point by living in a chicken coop. Regardless, David is among the sharpest people I know. He's also one of my mentors, and I often run big decisions by him.

"You just like to ask my advice so you can do the opposite," he jokes.

"That's only true half the time," I correct him.

David and I have a lot to catch up on. I want to get his take on the drama within the March community. "I think we're over the worst of it. I can't blame marchers for being concerned about money. But we've been raising enough to keep things going, and most marchers now see that."

I brief him on the meeting in western Nebraska a few weeks ago, about our "decision" on managing the budget. "That seems to have satisfied the worry warts, as far as I can tell. The other conflict was this ridiculous power struggle, which hasn't surfaced lately. So maybe we're beyond that, too."

A small cluster of businesses and homes comprise the center of Treynor, population 900, which sits on a hill with compelling views in both directions. To the east is the park where we'll camp tonight. After setting up our tents, David and I walk back into town where he hopes to enjoy a beer and I hope to find a signal to broadcast my program. We're successful on both accounts.

Returning to camp, I'm not overly surprised to learn there's a meeting in progress. Marchers have called so many meetings — and accomplished so little with them — that I've grown meeting-ambivalent. But David wants to check it out, so I pull up a chair and he sits on the grass about ten feet away. Lala's talking, and I settle in as she says, "He's not a bad person, he just does bad things."

"Hmm. Who's she going after now?" I wonder.

After a full minute of Lala rattling off a long list of undesirable qualities I realize she's talking about me. I cast an annoyed glance at David as if to say, "Well, looks like we're not over this shit after all."

Most marchers and a few guests are present. I'm embarrassed that, once again, visitors have to witness this internal squabbling. I'm angry that Lala has chosen to unleash a new attack on me during our first day in Iowa, tarnishing a pleasant and long-awaited homecoming.

After making me out to be the demented step-son of the King of the Underworld, Lala proposes that marchers elect a new board and vote me out as director. Jimmy, Kim, and Liz pile on, reiterating, with variations, Lala's list of my failures and personality disorders. I have to wonder how much time and effort the four of them spent planning this coup, and what they could have accomplished for climate action had their efforts been more productively directed.

As the attack escalates, I seethe. But a professorial glare from Osterberg reminds me to respond calmly. When it's my turn to speak, I take a deep breath and offer a rational dismissal of each point to occasional nods of agreement from other marchers. As with most March meetings, this one ends inconclusively, but with an understanding that there'll soon be another meeting to vote on Lala's proposal.

Marchers retire to their tents or private conversations. I'm tempted to debrief the meeting with Osterberg, but we opt to discuss it on the road tomorrow. "I'm really glad you're here, David," I tell him. "Probably couldn't have timed it better."

"We'll see," he says. "We'll talk tomorrow. Just remember, Fallon, don't be stupid!"

I laugh heartily. "Who, me? Stupid? You heard the entire catalogue of my faults just now — a long list, but 'stupid' wasn't on it."

"We'll see," concludes David. "Get a good night's sleep and your prospects for being stupid might diminish."

The next day, David and I set out with others for the 16-mile hike to Oakland. A couple hours into the day, David and I get a chance to talk privately. "This Lala, she's simply on a power trip," David observes. "She's used to being in control back home. She bosses her family around, decides what's for dinner, commands someone to do whatever chore she wants accomplished. She can't stand not being in charge here."

"You're not telling me anything I don't know," I respond. "And she's crafty, too."

"Yes, very clever, very conniving," continues David. "She waited to attack you when you were most vulnerable, here in Iowa. You need to be at the next meeting. Stay calm, be diplomatic. But most important, before that meeting, talk with friendly marchers. Enlist their support. Commit them to voting against Lala's proposal."

"Damn, you really love politics, don't you?" I observe. "I'm just not into these kinds of games. Besides, what Lala wants to do isn't even allowed under our structure."

"That doesn't matter. Play along with her game or you'll have even more rebellion. Just remember what's at stake," chastened David. "You've got an important mission and a message that people need to hear. Don't let Lala's power lust get in the way. If you ignore it, she wins, the planet loses. So, don't be stupid."

Occasionally, Osterberg gives me advice on personal matters, too, often unsolicited. He asks about my love life, and I tell him about Grace.

"Does she have money, Fallon?"

"What the hell does that have to do with love?" I ask.

"Does she have a job?"

"Part-time. So what?"

"Fallon, like me, you'll never have a lot of money. In fact, you have a lot less than me and you always will. So, don't get all silly and romantic. Crazy, young, pretty. That's what you've been after, right? And that's fine when you're 22. But you're an old guy now. So knock it off and find a woman your age who's not nuts and has money."

David continues his rant, "How much are you earning during this March?"

"Um, nothing," I say. "Actually, less than nothing."

"Right. So, hook up with a rich woman."

"Thanks, but this is one of those times I won't take your advice," I say. "Congratulations on batting .500 today."

The next day's march is 19 miles to Lewis. I feel sick as we set out and worse with each passing mile. I've had intestinal issues on the March, but this is different. It feels like I'm coming down with the flu. I move slower and slower as the day wears on, arriving at camp chilled and feverish.

After a worthless night's sleep, I'm more debilitated than ever, even worse than when I staggered into Wiggins. The short walk from my tent to the Eco Commodes is difficult and leaves me dizzy. Today's march is just nine miles and tomorrow we have the day off. So I opt to focus on recovery, hoping that a two-day break will have me back to normal and able to catch up.

Osterberg rents a hotel room outside Atlantic. He brings me Tylenol and some crappy food, which I barely touch. I hardly get out of bed for two days and nights, sleeping most of the time.

On August 5, I'm still under the weather, but the fever's gone and I feel well enough to attempt the 13-mile march from Lewis to Cumberland. Osterberg accommodates my slower pace and we arrive at camp late afternoon. I'm grateful to my friend and mentor's help and patience, grateful that the string of footsteps that began on the Pacific Ocean didn't have to end less than a week from my home.

In Bridgewater, population 173, we camp in the city park. Residents organize a potluck and we enjoy an evening of genuine Midwestern hospitality. But Bridgewater is uncharacteristically noisy, with every dog in town taking a 20-minute shift to bark incessantly. To add to our auditory torment, some guy in an open garage across from the park runs his power saw until midnight.

The next morning, we discover a cafe where some of us enjoy a robust breakfast. Our route today is along quiet roads, many of them gravel through some of the most beautiful countryside western Iowa has to offer, dotted with fields of towering wind turbines. We arrive in Greenfield, the seat of Adair County and famous for Vice President Henry Wallace, the movie *Cold Turkey*, and the state headquarters of Jesse Jackson's 1988 campaign for president.

But the day is soured when an afternoon meeting is called to vote on Lala's proposal to fire me. My skewering continues with renewed vigor. Kim mocks me with accusations of being a slick politician, and I lean over to Osterberg and say, "Ha! Don't you wish!"

Lala announces that if I don't accept the results of the vote, she wants marchers to boycott the events in Des Moines. She even threatens a hunger strike. Several marchers speak in my defense. The tone of the meeting is ugly and contentious, the worst yet, and ends inconclusively with marchers agreeing to reconvene and vote after supper.

"Well done on maintaining your composure," says Osterberg.

"Thanks, but this shit just makes my stomach churn. Between the stress and lack of sleep, I'm inviting a relapse of that nasty flu. These attacks by other marchers, two who I used to consider friends, are honestly worse than anything I ever had to deal with as a legislator," I say to David. "I've gotta get outta here tonight. I'm skipping the meeting and Lala's meaningless vote and gonna stay at the hotel in Greenfield. Wanna join me?"

"You should stay for the meeting," says Osterberg. "But yeah, we can't have you getting sick again."

David pauses. "How many friendly marchers have you talked to these past couple of days?"

"Lots," I reply. "A dozen. More than a dozen. And they're solid."

"Good. You've made your case, and it looks like you've got plenty of allies. So, you go stay at the hotel. I'll stick around for the meeting and let you know how it goes."

I thank David, catch a ride to Greenfield, and check into the town's lone hotel — a historic building on the square that was recently renovated. I send a message to Grace. Greenfield is only an hour from Des Moines, and I hope she might be able to visit. Months ago, I suggested we meet up in Greenfield, but she remained noncommittal. She responds to my message in just a few words, saying she has other plans and can't talk. She'll see me in two days when we camp just south of Des Moines at Sarah's family's farm. I try not to read too much into her message, but I detect reluctance.

I have supper at the hotel by myself. The food is less than gratifying, and I'm reminded that any meal prominently featuring iceberg lettuce is doomed to failure. The evening improves when I discover the hotel's electric baby grand piano, which sounds remarkably better than most electric pianos. I play a few nocturnes with some satisfaction and look forward to being home with my piano soon.

The next morning, I walk to The Corner — a cafe on the town square where I enjoy a hearty breakfast of sausage panini with fresh-baked scones and a hot cup of tea. I find a ride back to camp for the 14-mile march to Winterset. Marchers have already left when I arrive, but I catch up on the first break. David apprises me of last night's meeting.

"I know you're glad you missed it, but you would've enjoyed the expression on Lala's face when marchers soundly defeated her proposal to fire you," Osterberg says, beaming. "She was genuinely shocked. She only got five votes."

"Well, that's a relief," I say. "See what a good mentor you are?"

"Maybe, but I'll be gone in few days, and you've got more politicking to do," warns David. "Lala's not going to quit. And while most marchers are happy to keep you on as director, a lot of them feel changing the composition of the board makes sense. So, be prepared for that to come up. You aren't out of the woods yet, Fallon."

I think about that as we walk and begin to formulate a plan. I suggest adding a couple marchers to the existing board. They could be in charge of approving the plan to disperse any additional funds after we've finished the March.

"Good! Now you're thinking like a politician," says David proudly, though I'm reluctant to regard this as a compliment.

In Winterset, local supporters organize a potluck dinner and community dialogue at the public library. After a wonderful burrito-themed supper, Robert Knuth, a Lakota man, tells us he held a blessing ceremony the day the March set out from Los Angeles. He hands me a bundle of sage, sweet grass, and cedar. He also hands me a walking stick and a sash that belonged to his brother, who passed away recently.

"The black and white sections of the sash represent humanity," explains Robert. "The blue and green parts represent the Earth and Sky. I want you to have these, Ed, because what you and these other marchers are doing is critically important. You're carrying the hopes and dreams of all humanity with you."

It's a powerful moment. Several marchers are visibly moved. Robert says he'll hold another blessing ceremony for us the day we arrive in Washington, DC.

Later, I reflect on Robert's gifts, remembering all that we've received from Indigenous people during the March — the drum ceremony in Los

Angeles, the turquoise stones given to Miriam and Steve in Phoenix, the continuous hospitality and guidance during our time in New Mexico. Now, these gifts from Robert. Such powerful medicine!

The memories capture my heart as I sit outside my tent on the high school's football field, fireflies dancing all around me. To the north, the stately Madison County courthouse illuminates a darkening sky. To the west, the red and orange glow of a mid-summer's sunset spreads across half the horizon. To the east, the silhouettes of hunting nighthawks dart in front of the rising moon. To the south, way off in the distance, streaks of lightning flash among towering billows of cumulus grandeur. I feel both humbled and embraced, surrounded on all sides by this multi-dimensional light show, most of it untampered with by man, all of it glorifying life, light, and creation.

Dew forms on the grass, coating my tent, moistening my bare feet. There are no insects to drive me inside so I sit and absorb the lessons of the night. I reflect upon the future of these five points of light in the New Climate Era. Sun, moon, and lightning will glorify Earth's skies long after humanity has vanished. But the fate of fireflies hangs in the balance, as does the immediate and long-term future of Winterset's town square. Even with all its flaws and the accompanying slaughter of the bison and attempted genocide of Native peoples, what European settlers built here on this great prairie is a thing of beauty worthy of reverence, respect, and preservation.

I think of Robert's words: "You're carrying the hopes and dreams of all humanity with you." True enough. If the March and all who are pushing back against climate change fail, humanity, not Earth, will be the victim. Earth will get along just fine without us.

Though we're called "One Earth Village," the March is not so much about Earth as it is about us, about our species and the millions of other species whose continued existence is in doubt because of human greed and ignorance. Can we summon the wisdom and courage to change? Can we learn quickly to live in harmony with the natural order? Or will we continue to destroy and defile the great gifts Earth has given us?

Sitting in this field, marveling at Earth's brilliant light show, I feel at peace even as I know the clock is running out on humanity's game of climate jeopardy. The questions have been posed. Most of us understand them. Our answers will have to come soon, if it's not too late already.

CHAPTER 28

GRACE

"The elusive nature of love... it can be such a fleeting thing. You see it there and it's just fluttering and it's gone."

— MICK JAGGER

Our 20-mile march from Winterset to the Spain family farm goes by quickly. I'm excited to be so close to home, eager to see some of the friends and acquaintances I've missed. Yet most of my thoughts during today's eight-hour march are about Grace. In May, I was certain we'd meet in Greenfield or Winterset and possibly spend a week or more together during my time in central Iowa. Yet through Colorado and Nebraska, our phone conversations became shorter and less frequent. The closer I got to Iowa, the more distant she seemed. I never allowed myself to believe anything was wrong, but now I'm uneasy. I try to reassure myself that she's simply been busy and everything will be fine.

The last stretch of today's march is on gravel. When a car or farm vehicle passes, we inhale occasional gulps of dust. But this is a small inconvenience given the break from traffic. Sarah and her family have organized a grand reception, with a bonfire, luminaries, music, reiki, and a potluck featuring food from nearby farms. Over a hundred local supporters are expected to join us.

I sit in front of my tent hoping to catch up on phone calls and email before guests arrive. I'm on a call when Lala and a few other marchers show up unexpectedly and plop themselves cross-legged on the ground in front of me. Feeling pressured, I rush through my call and ask, "What's up?"

"We want a meeting immediately, as in right now," demands Lala. "We need to elect a new board, and you keep putting it off. Let's just do it."

"Are you for real?" I say curtly, as more marchers arrive. "Do you realize how rude it is to pull a stunt like this while Sarah and her family are running about trying to get everything set up for tonight?"

"Marchers might have voted to keep you on as director," says Lala. "But they also want a new board and they want it now."

"Look, we can vote on whether to have a new board soon," I insist. "But you can't demand a meeting out of the blue. I've got an open mind on the make-up of the board. But we have to give all marchers ample notice. Let's schedule it for Grinnell."

Lala is marginally placated and she and the others leave. I close my eyes for a minute and try to calm down. But I'm shaken and decide to put off the rest of my work until tomorrow. I help Sarah set up chairs and tables, glancing down the gravel road every time I hear a car approach, eager to greet whoever shows up, imagining that each vehicle nosing its way over the road's crest is Grace's.

To bless the fellowship and food, Sarah and I ask marchers and guests to gather around the fire. Sarah sings a Navajo blessing song. In a large shell we burn the cedar, sweet grass, and sage that Robert gave us. Sarah and I walk around the circle with the smoldering plants as people smudge themselves with the sacred smoke in a moment of prayer, reflection, and gratitude.

Grace arrives as people line up for dinner. I hurry over to her car, feeling like a man who's wandered through a wasteland to finally stumble upon a long-desired oasis. Grace smiles when she sees me, and I want to hold her tight, feel her melt into my parched arms for a hug that quenches six months of emotional thirst. In my mind, I envision an embrace like the one we shared when we danced together in my living room over a year ago.

But her greeting is brief, a one-armed hug that fades quickly — a mirage, not an oasis. A friend has come with her so there's no time to talk. We walk toward the barn where marchers and guests are mingling. Grace tells me she can't stay long. She's got a lot on her plate during the next two days, but she'll try to break away and visit me for a bit while I'm in Des Moines.

I feel a vast, empty pit form in my gut as we join the line for dinner. This is not the reunion I expected, dreamed of, longed for. The potluck's offerings are a delightful sampling of summer's abundance, but nothing tastes good to me.

Later, away from the crowd, Grace and I find a few minutes to talk. I ask if she'll be able to come tomorrow for the last two miles of our march to the State Capitol. She tells me she has another commitment. I point out that this entire week's activities have been on the calendar since January. She gets annoyed, tells me I'm too persistent, that I'm not understanding the demands on her time and schedule.

We pause for an uncomfortable moment. I ask her point-blank if she still wants to be with me when the March is over. She says she's not sure. She's worried about health insurance, about whether I'll have a job and an income, about other details that to me seem trivial for two people in love. I tell her I think I've deluded myself into believing that our love is mutual. She assures me it is, but I know in my heart it isn't. I've been fooling myself, but I don't understand what went wrong, don't know why I haven't seen more clearly that we're in two entirely different places.

Grace says goodbye. She waves awkwardly as she gets into her car. I wave back. Her car door clicking shut strikes me as the saddest sound I've ever heard. I stare at the dust kicking up on the road as her car disappears over the hill in the distance.

I wander back to the party dazed, doing the best I can to mask my sadness. I visit with friends and acquaintances who tell me how grateful they are for what we're doing. They're eager to hear about the experience. I share anecdotes of how we've connected with people who hadn't given much thought to climate change. But my mind and heart are elsewhere. It's as if I'm sitting in the balcony of a theater observing myself, watching the gestures and lines of an actor who knows what he must say but has lost interest in both his audience and the performance itself.

Maybe some of my friends sense my detachment. A few remark how thin I look. I tell them I've lost weight, but I say nothing about having just lost so much more, having lost the love of my life.

That night, after guests and friends leave and most marchers are asleep, I compose my farewell letter to Grace under the light of my headlamp:

> Dearest Grace,
>
> I love you so much. For over a year, I've wanted to be with you, wanted to share my life with you. But clearly, I've been deluding myself, imagining that you love me as deeply as I love you.
>
> One measure of love is time. When you love someone, you make time for them. You prioritize them. You go out of your way to be with them, as I have done on so many occasions to spend even twenty minutes with you.
>
> There have been occasions when I felt similarly important to you, like in February when you helped me find clothing and camping gear for the March, or when you made me that beautiful locket. Mostly though, I feel like a sidebar, an extra, someone you call when you're doing something else (usually driving). I'm the activity you squeeze in between two things that are more important. This has been the case for most of our relationship, and I see that more clearly now, with the most flagrant example your unwillingness to carve out time for us during my days here in central Iowa.
>
> It's obvious to me you don't feel the same as I do. I need to accept that and move on. For several weeks, I've suspected this was the case but didn't want to admit it. I kept hoping for some indication that I was wrong, some sign that you loved me and wanted to be with me. This week's blow-off is solid confirmation that our hearts are in different places.
>
> I've carried your locket every step of the way since Los Angeles. It reminded me of your love and gave me hope for

our future together. At night, I'd move it out of my pants pocket into the pocket of my tent. Now, with great sadness, I'll leave it in Des Moines when the March heads east.

Perhaps this is the end of the conversation, the conversation we never really had. If something changes, if you decide you want to be with me, you can always call or write. But I can't promise I'll be available.

Time may be the measure of love, but it's also the enemy of life. As my time on Earth lessens with every passing step, life beckons me to find a companion to share its joys and sorrows. Perhaps there's a woman out there who'll love me as much as I'm ready to love her. I guess I'll throw my heart open to the world and find out.

It's beyond hard to write these words because I was certain you were the one I was supposed to spend the rest of my life with. I feel like a fool. I'll always love you, and I miss you more and more with each word I write. Yours with an aching heart,

Ed

A dozen or so friends join us for the 18-mile march from the Spain farm to the Iowa State Capitol. I put last night's sadness behind me and share in the exuberance of what for several Iowa marchers is a homecoming. This is Osterberg's final day on the March. I tell him about last night's conversation with Grace. "Well, the love of my life didn't pan out."

"Good. Glad to hear it," says David.

"You really are a heartless son of a bitch, aren't you," I say.

"No, I'm just older and wiser than you, and I know you need financial stability in your life if you're going to continue your work, which you need to do. People are counting on you."

"In that case, I'm counting on you to send a beautiful rich woman my way soon, cause I'm sick of this monk's life."

"I'll see what I can do," says David, though we both know he's not serious.

"Hey," I say to David. "I can't tell you how much it means that you joined the March for two weeks. You really helped me out during a tough spell."

David smiles. We're close to the outskirts of the city and our conversation pauses as we negotiate a stretch of road with increased traffic and a crossing unfriendly to pedestrians.

"I gotta tell you, Osterberg, I just wanna stop here in Des Moines, call it good. Los Angeles to Des Moines: That's enough marching for one guy to do for the planet, isn't it?"

"Sounds like you're asking for advice, so I'll take a shot at it," David responds. "You've done more for the Earth this past year than most people do in a lifetime. You could quit now and no one would fault you. But you made a commitment, and if you're at all physically and emotionally able to sustain that commitment, you need to keep going. Besides, if you quit, you'll make Lala happy, and you don't want to do that, do you?"

I laugh a bit. "Vindictiveness is no reason to keep marching. But yeah, commitment. That's a big deal. I'll think about it as we march."

The downtown Des Moines skyline and the State Capitol are now clearly visible. "There's our former office," I say to David, admiring the sun's glint off the golden dome of the Statehouse. "You walked to your old office all the way from Nebraska. How about that?"

"And you from California," says David.

I think about that as we merge onto a busier road. I've marched 2,000 miles from the Pacific Ocean to my home in the Heartland. Maybe that's enough, I tell myself. I imagined that our month across Iowa would be a respite from the rigors of the March. Instead, I've had to wrestle with my worst bout of illness yet, endured three hostile meetings, and now the sudden end of my relationship with the woman I love. Maybe these are signs that, for me at least, the March is over.

Yes, that makes a lot of sense. I let this idea sink in as my apartment, cat, and piano grow closer. The more I dwell on it, the more right it feels. This is it — the grand finale, the final stretch. I'll use my speech on the steps of the Iowa Statehouse tonight to tell people that I've gone as far as I need to go, that other marchers will plow ahead without me, carrying the message of climate urgency to President Obama. Besides, with the Dakota Access pipeline now a growing threat, my work is here in Iowa.

"Yes, this is a good plan," I say to myself and smile.

CHAPTER 29

SON AND FATHER

"When a father gives to his son, both laugh; when a son gives to his father, both cry."

— JEWISH PROVERB

In the fall of 1985, Benjamin E. Fallon was born to an eager mom and a clueless dad. You've heard touching stories of men overcome with emotion at the birth of their child, for whom the experience was profoundly transformational. That wasn't me. I was overcome for sure — by the blood, the screaming, the exhaustion. My first thoughts upon seeing Ben emerge into world were, "Holy shit! Really? That's what they look like? Why is his head flat? Won't that mess up his brain?"

It was as if I'd made contact with an alien.

For the first few weeks, Kristin provided just under 99.9 percent of the care. Then the dreaded moment arrived: *she left for two whole hours!* I was alone with this strange, irrational creature. "Please let him sleep until she gets home," I begged. "Please let him sleep."

Ben woke shortly after Kristin left. I picked him up. He cried. I rocked him. He screamed. I carried him around the house, holding him more like you'd grasp a hot loaf of badly burnt bread than a human child. He wouldn't stop crying. Nothing I did made a difference.

When Kristin finally arrived home I was in full panic mode. "He hates me," I said. "I swear, he hates me. I fed him. I changed him. I walked around and around trying to calm him down. Then I accidentally banged his head on the wall and he cried even more. I hope I didn't do any permanent damage."

Fortunately, my stint as a lousy parent was short lived. I'd started to learn Spanish two months before Ben was born. To improve my own skills and expose Ben to a second language, I spoke only Spanish to him. We sang songs and read books in Spanish. We invented games involving animals, food, body parts — all in Spanish.

As Ben got older and more mischievous, Spanish became our secret code talk. Sometimes, I'd tease Kristin in Spanish and Ben would laugh as we'd wait to see if she understood. "Hey, are you talking about me?" Kristin would ask. Ben would laugh even harder, and we'd often continue teasing her even though we'd been outed.

Besides the fun of being able to talk about others behind their backs, there are good reasons to raise a child in a bilingual household. For me, the most tangible benefit was that it facilitated the critical father-son bond. What began as an effort to learn and teach a foreign language became the life jacket that saved a struggling new dad who'd never been given an operator's manual on how to navigate the choppy waters of parenthood.

I relish the March's slow trek to Des Moines, broken up by numerous stops, some planned, some spontaneous. Toward late afternoon, we arrive at Ritual Cafe where volunteers serve Ben and Jerry's ice cream to a crowd of supporters assembling for the final two-mile march to the Statehouse.

On the sidewalk in front of the cafe I reconnect with more friends than I ever imagined would show up. It's a festive gathering, like Christmas with just the family members you actually like rolled in with a class reunion where nobody gets drunk or looks old.

I hug and talk with one friend after another. Troy Church, who's staying at my apartment, is telling me how much he dislikes every cat in the world except Mika when suddenly, in the most unexpected moment during a year in which unexpected has been normal, my son Ben slowly drives by with his dog Diggy, heads hanging out the windows, tongues hanging out of their heads.

I'm in a state of blissful shock. I run up to the car. "What the hell are you doing here? Did you drive all the way from Florida? How come you didn't tell me you were coming?" I rattle off one question after another before Ben has a chance to answer any of them.

With a huge smile, Ben says, "You want a couple more marchers for a few days? Diggy's chomping at the bit to check out the Iowa countryside and chase a few squinnies."[1]

My eyes moisten as I give Ben the biggest hug I've ever given him. "Don't surprise me like that, you brat," I say as I fight hard to choke back tears.

I lean against the car and look at this boy-turned-man, giving thanks that the time I banged his head against the wall seems not to have caused any cerebral disfunction. I ask, "Really! You've come all this way to march with me?"

"Yeah, if I can't keep up with my old man, what good am I?"

Ben and Diggy join nearly 400 people for the largest march and rally we've had since Los Angeles. I have little time to think about revising my speech, but I know my son didn't drive 1,500 miles to see his dad quit.

At the rally, I tell everyone how grateful I am that such an impressive crowd has turned out to support us. I remind people that our work as marchers isn't done, and in two days we'll set off to finish the longest climate march ever.

"We're gonna persevere to the end," I yell into the mic. "And we're counting on *you* to persevere as well, not just through the end of the March, but until our elected officials and all Americans embrace the absolute urgency of climate change."

I don't say it in my speech because I know it would bring a flood of tears to my eyes. But in my heart I thank Ben for giving me the best gift a son could give: hope, courage, and a willingness to walk through the most challenging of times, so challenging that a couple hours earlier, I was ready to give up.

---

[1] The name is unique to the Des Moines area. A "squinny" is what biologists call the 13-lined ground squirrel.

CHAPTER 30
MY PIANO

"Sometimes I can only groan, suffer, and pour out my despair at the piano!"

— FRÉDÉRIC CHOPIN

After the rally, I catch a ride to my apartment with Troy, who offers to camp out so I can sleep in my own bed for two nights. It's a strange feeling to turn the lock on the back door after having been away for nearly six months. My first thought is, "I'll bet my cat will be happy to see me."

I walk through the apartment in search of Mika and find her sleeping on a chair, curled up in a furry ball of feline contentment. She looks at me for a couple seconds, then puts her head down to go back to sleep, as if to say, "Nope, haven't missed you a bit, silly human. Next time you abandon me, don't let the door hit you in the ass on the way out."

I walk slowly around my apartment, scanning the interior, noticing what's the same and what's different. I feel oddly anxious and don't understand why. Two weeks ago when we crossed the Missouri River, I felt overwhelmed by a sense of excitement at coming home. That sense grew stronger as we drew closer to Des Moines, stronger still during today's march and rally. So why this sudden, unexpected anxiety?

I walk over to my piano, a 100-year-old Mason & Hamlin grand, the one possession I truly treasure. Approaching the keys with reverence, I begin to play Chopin's Opus 27 #1 Nocturne — "The Great Nocturne" as my eccentric college piano instructor, Nigel Cox, called it.

With my piano, every piece is a duet — my hands and her keys — creating waves that are more than just sound. The melody of the nocturne's opening section enters dark and moody, with singular, slow-moving notes painting a portrait of human sadness and solitude. I feel both relief and suspense when a muddied, two-measure interlude drags us into the nocturne's blustery midsection. Like a string of strong summer storms, my piano and I flirt with distant rumblings that grow closer and more ominous.

Suddenly the skies burst — but not with rain, with sunshine! Filtered streaks of light glint through the thunderheads in a statement of celestial grandeur even more powerful than the storm itself. But the clouds soon return, filling in the gaps where the sun dared poke through. Finally, the full brunt of the storm strikes, then relents, then strikes again until bass octaves roll off into the distance, bringing us back to the opening melody.

But what's this? The melody is no longer melancholy! Transformed by the storm, a passage that earlier resonated with darkness now reigns triumphant, dignified, accomplished. Same notes, different meaning. The Great Nocturne resolves into a lilting barcarolle, gently rocking me and my piano forward and back, floating in a birchbark canoe across water as calm as glass, across a pond bejeweled with just the right amount of setting sunlight, a day ending in perfection, peace, and contentment.

As occasionally happens when I play Chopin's nocturnes, especially after an extended hiatus, life's rough spots come into focus. I now understand why I feel anxious. I've grown accustomed to the freedom of a life lived largely outdoors where the walls that separate me from the world are thin and mobile. Here, I'm penned in by "civilized" walls, barriers of wood and brick that insulate me from cold and storm, yet also isolate me from people, nature, and life.

Over the past six months I've become wilder, freer, less fit for modern living. Despite frequent and intense conflicts with marchers, they are my tribe. Though I often walk alone during the day, at night I return to a village where we're physically connected in a chaotic but comforting cluster of tents and vehicles, where our communal center of fire, food, and water is equally accessible to all. It's weird, disturbing even, to admit that I miss this dysfunctional community I'm often at war with. But tonight, that is The Great Nocturne's lesson. I listen, embracing its uncomfortable wisdom.

There's another reason for the anxiety I feel in my apartment, of course — the realization that I won't be sharing this space with Grace when the March is over. Even though I've broken up with her, I can't let go. I spend the day off in Des Moines mostly by myself, enjoying some downtime but also raising money and troubleshooting questions about route, rallies, and home stays in the coming weeks. It's hard to focus on work because I keep thinking Grace will stop by. She doesn't.

The next morning comes quickly, and marchers assemble at the State Capitol at 7:00 for the 13-mile trek to Mitchellville. I can't abandon the delusion that Grace might visit, so instead of departing with the others this morning, I linger at my apartment, waiting and hoping.

As a result, Ben, Diggy, and I don't leave Des Moines until 4:00 p.m. We arrive at the March camp well after dark on a pitch-black night where walking on the road becomes dangerous. I feel stupid for starting so late, stupid for keeping Ben waiting and exposing him and Diggy to these conditions, stupid for believing that Grace would come see me.

Over the next two days, the March covers 40 miles, arriving in Grinnell mid-afternoon on August 15. Because of our late start out of Des Moines, it's the longest distance I've covered in a 48-hour period — 53 miles! I'm amazed that Ben and his dog are able to manage it. Diggy is exhausted, but Ben shows no sign of fatigue and never complains.

That evening at our campsite in Grinnell, marchers meet to discuss my proposal on what to do with the board of directors. The meeting was scheduled with ample notice and publicized with an agenda that everyone knows about well in advance, a fact that I make sure to mention. I inform marchers that I've talked about their concerns with Shari and other key people, and they agree that adding two marchers to the board with full voting rights is a good idea. This is more than Lala imagined I'd accept, and the proposal sails through with little discussion. Even though the expanded board does nothing to alter the power structure within the March community or within the organization itself, marchers are satisfied with the compromise and pleased that the meeting was brief, efficient, and actually accomplished something.

I use this opportunity to leverage renewed commitment to focus our efforts outward on the mission of the March. Heads nod in agreement, but I'm skeptical as to how long it will hold up, wondering if we'll even make it through Illinois without some other major March malfunction.

Yesterday morning, Grace wrote a one-sentence message to say that my letter was the most painful she'd ever received. Last night, she called and we spoke for 20 minutes or so, mostly small talk. She said she would try to come to Grinnell to see me as it was such a short drive from her home. I'm doubtful, but it's impossible not to get my hopes up.

Grace doesn't come to Grinnell. I'm disappointed, but also relieved. It's been an excellent day. I say goodbye to Ben, reminding him that his timing couldn't have been better. I can't imagine how horrible I would have felt had I quit in Des Moines.

CHAPTER 31

HOME

"A man travels the world over in search of what he needs and returns home to find it."

— GEORGE A. MOORE, *THE BROOK KERITH*

Our trek across eastern Iowa is everything I'd hoped all of Iowa would be. The trifecta of misery that afflicted me from Council Bluffs to Des Moines — microbes, mutiny, and heartbreak — are temporarily forgotten in the friendly encounters and internal harmony the March enjoys from Des Moines to Davenport.

Frequently, local supporters host us with a potluck dinner. In Newton, Troy's father, Ted Church, entertains us with his accordion band during supper. Near Ladora, we pass a man sitting at a table along the highway giving out water. He's crafted a handmade sign reading, "For Friends of Fallon."

Drivers often stop, and nearly all are positive or at least friendly. Once, walking with Kathe Thompson, a 72-year old retired school teacher from Florida who's been on the March since Los Angeles, a man rolls down his window, curious about who we are. It's Andy McKean, a Republican lawmaker from Anamosa. He assures Kathe and me that, while we might disagree on a lot of things, he concurs that climate change is a major challenge we need to address.

"My, do you know everyone in Iowa?" Kathe asks.

"No," I respond with a chuckle, "maybe half the population, but not everyone — at least not yet. But seriously, it does feel good to be home."

On August 18, we arrive at the Amana Colonies, a cluster of seven villages founded in 1856 by Pietists fleeing religious persecution in Germany. The villages were managed communally for 70 years until the Great Depression forced a change in their governance structure. Private ownership of homes and businesses was allowed, but communal ownership of the land continued. That helped Amana avoid the inevitable inundation of chain businesses that now dominate most retail economies. Amana's bakeries, restaurants, breweries, wineries, furniture shops, and the state's only woolen mill now attract close to one million visitors each year.

We pitch our tents in front of a grocery store where I chat with the owner of a nearby craft shop — a witty, gregarious fellow who's happy to share his thoughts on the economy. "People visit us because they love our quaint buildings and pretty countryside, which isn't cluttered with billboards or hog confinements like a lot of places. Everything's unique and local, and that draws folks in like cattle to a shade tree on a hot summer day."

The man jokes as he continues, "Basically, I make my living off nostalgia. "If every town looked like Amana, I'd go out of business. A few years back not far from here, a bunch of slick East Coast guys in suits showed up and convinced city hall the town would dry up and die without chain stores. So city officials gave them everything they wanted — free land, tax breaks, even a new road. What happened a few years later? Sure enough, Main Street dried up and died. Those bastards killed the very heart of the town they said they'd come to save."

"That sounds familiar," I say. I share one of the many tales from my work as a state lawmaker. "Once, out of the blue, a corporation in western Iowa threatened to move out of state if it didn't get incentives. The Legislature put everything — I mean *everything* — on hold for a solid week while it figured out how to hand over millions to the company. A few years later, the company threatened to leave again."

"Well, it's easy to point the finger at politicians," the man says. "I'm happy to do that, but ultimately it's the people's fault. In the case of the town I was talking about, *people* elected those bozos who sold-out to the big chains. What's worse, those same people chose to shop at the chains instead of on Main Street. They thought it'd save 'em a couple bucks. But next thing you know, their taxes got raised to tear down the abandoned buildings on Main Street. The last person to own a business there moved out two years ago. He's now a greeter at Walmart."

The guy's on a roll, but he stops as we both ponder whether to laugh or cry. "Don't tell anyone," he says with a glint of mischief in his eyes. "Don't let 'em know that if they had the balls and the guts their town could be just as lively as Amana. 'Cause if they figure it out, what we've got here won't be special anymore. I'd have to close up shop, probably have to get a job as a Walmart greeter," he says laughing.

I think about how, like the Amana Colonies, the March is attractive not merely as a passing curiosity but because it taps into the human hunger for intimacy with people, place, and history. I think back to what Bill McKibben said at Iowa State University seven years ago when my journey as a climate warrior began: "Americans will be happier if they return to the 1950s lifestyle of eating together as a family, talking with neighbors, and carpooling to work."

Weaning America from fossil fuels is an enormous challenge. Convincing Americans to step out from behind the walls, screens, and built-in isolation that's come to feel normal may be an even bigger challenge.

Leaving the Colonies, we march through the antithesis of Amana, past the numbing onslaught of Iowa City's commercial sprawl. The following day — our "day off" — begins for me at 3:30 a.m. with a local TV station conducting five interviews over the course of three hours. Afterward, I join marchers, along with a big crowd of local supporters, for breakfast and a three-mile parade into downtown Iowa City for our rally.

Iowa City is Miriam's town, and she's thrilled to be home for a couple days. At the rally she steps up to the mic as her friends drape a green super-hero cape around her shoulders. Like the rest of us, they're in awe that Miriam's walked every step of the way.

I think about how difficult this march has been on my body, then add fifteen years and try to imagine what Miriam has been through. I regret the month when we didn't getting along. But we moved beyond that and now feel bonded by our common purpose, by the importance of our footsteps in this life-and-death struggle to mobilize America to fight back against climate change.

In her speech Miriam thanks people for their support and encouragement and says, "If Americans would invest just half the time, attention, energy, emotion, and money into solving the climate crisis as we do following sports, we could solve this in an amazingly short time. America geared up quickly after Japan attacked Pearl Harbor. We can do it again. We just need to face the truth that is staring us in the face and rally ourselves. No one is going to do it for us."

As Miriam addresses the crowd, it seems to me her super-hero cape makes her look smaller, not bolder. I've known Miriam for less than a year, but I can't help notice how much older she looks, how thin and weathered. How could she not? What she's done — walking over 2,000 miles with only an occasional day off — is medically ill-advised for someone her age, and well, the sort of accomplishment you'd expect of a super hero.

Months ago, marchers elected Miriam the Mayor of One Earth Village. Today, she's the Warrior Queen come home to her people, resting for a day before helping to lead our ragtag assemblage of renegade patriots into battle for the final thousand-mile crusade.

Our walk from Iowa City to West Branch is quick and pleasant. I arrive at Scattergood Friends School just in time for a recording session with the Climate Justice Gypsy Band. We're publishing three songs we hope will generate some revenue. We sound better than we did a couple months ago but remain as musically undisciplined as the March itself. Still, thanks to decent acoustics and some professional guidance, our recording doesn't come off too badly.

The following day we walk 17 miles to Moscow through the beautiful, wooded flood plain along the Cedar River. Our campsite along an oily gravel road presents an ugly contrast to the river. The oil was applied to keep the dust down — hardly an ecologically advisable practice. But this beats all. Never have I seen so much oil on a road. It reeks and is made worse by last night's rain, with oil coating the puddles of standing water. For the sake of state pride, I'd hoped every campsite in Iowa would rank in the top 50 percent. This one ranks near the bottom.

Our 18-mile march from Moscow to Walcott is made easier with the assistance of two cafes. My camp breakfast of oatmeal, pecans, blueberries, and yogurt holds me through Wilton. There, I dig into a second breakfast of eggs, sausage links, hash browns, and pancakes. At the next cafe in Durant, I order my third breakfast and marvel that, despite this voluminous consumption of bad food, I neither gain weight nor develop gastrointestinal problems. The novelty of these cafes wore off long ago. I'm now just eating to survive, following Charles's orders as I dutifully consume my pricey medicine of protein and calories. Cafes that used to exude charm and the alluring scent of bacon now reek of grease and heart failure. As I pan today's offerings, my mind wanders back to the oil-coated puddles along the gravel road in Moscow. "That's a different kind of oil, silly," I chastise myself. "Just shut up and eat your medicine."

On August 24, our final day in Iowa, we march 13 miles to Davenport through hot, humid conditions. I'm ticked off that only six climate marchers participate, outnumbered nearly two to one by locals. But I put that behind me. Supporters have gone to great lengths to organize a stellar greeting at LeClaire Park by the Mississippi River. We arrive on schedule for an impressive smorgasbord in an air-conditioned building. After lunch, we go back outside into the heat for our rally. We receive some of the best press coverage we've had on the March and Mayor Bill Gluba is one of the speakers.

Later, Bill and I catch up over a beer. He tells me how climate change is affecting the Mississippi River. "I remember a flood in 1965, then one in the mid-1980s. Now it's a recurring situation, nearly every year," he says.

Bill owns property just south of Davenport. "I've had that land since 1986. I planted trees on it, but with the frequent floods they've all died. I don't even bother to plant anything there anymore."

Today, the mighty river stays within its banks. It's a different river than it was just a couple decades ago. It's hard to understand why climate change isn't evident to everyone living along its banks. Today, the river speaks to me of the urgency of action to address the climate crisis. Tomorrow it will have another, more personal, message.

CHAPTER 32

THE GREAT RIVER

"The Mississippi River carries the mud of thirty states and two provinces 2,000 miles south to the delta and deposits 500 million tons of it there every year. The business of the Mississippi, which it will accomplish in time, is methodically to transport all of Illinois to the Gulf of Mexico."

— CHARLES KURALT

Twenty-thousand years ago, the dominant north-south flow of the Mississippi River was interrupted by an ambitious glacier near what is now Davenport, causing the river to flow east-west for about 40 miles. Because our first day's march through Illinois follows the left-descending bank of the Mississippi we never lose sight of Iowa. I relish what feels like a bonus day in my home state, thanking the glacier that made it possible.

We set out from LeClaire Park at 8:00 this morning, the mayor and a handful of other local supporters joining us for the walk across the Mississippi River to Illinois. Dressed in a suit coat and tie, Bill presents a stark contrast to casually and colorfully clad marchers.

Approaching the Arsenal Bridge, we're surprised to see two people high up, holding what looks like a large sign. "What's that all about, Bill?" I ask.

Before he can answer, our huge Climate March banner is unfurled and draped over the side of the bridge. Bill looks at me and says,

"That's not even remotely legal, you know. This bridge is controlled by the federal government, and officials are pretty firm in their response to this sort of thing."

"Honestly, Bill, I had no idea this was going to happen," I say. At the same time, I smile to consider how many commuters get to see our message.

We're now close enough that I can make out Mack and Sean holding either end of the banner. We arrive as two officers are talking with them. Rather than risk arrest, Mack and Sean agree to reel in the banner.

"Sorry about that, Bill," I say. "Hope that doesn't cause you any problems."

"Naw," says Bill. "Off the record, I'd say it was pretty cool." I'm reminded why Bill is one of my favorite local elected officials.

About halfway across the bridge dark clouds move in from the West. It seems doubtful we'll make it across before the storm hits, and I wonder how Bill's suit will hold up to what promises to be a solid drenching. But we move quickly, shuffling into the Rock Island Botanical Center as strong winds tear leaves off trees and a hard rain begins to fall.

Inside, we're treated to music, food, and a welcoming ceremony by Rock Island Mayor Dennis Pauley. Bill had given us a symbolic key to the City of Davenport, telling me, "I don't give many of these out. But you guys are doing such important work, it seems appropriate." Mayor Pauley now performs the same gesture of kindness, presenting us with a key to the City of Rock Island — one mayor passing us off into the care of another as we cross the iconic divide between America's west and east.

Across Iowa, I mostly walked with others. Today, I want to enjoy the river and final vistas of Iowa by myself. So I walk alone at a casual pace. There's work to accomplish, too, so I take a long lunch break in Moline, Illinois at the Barley and Rye Bistro. I also pop into Lagomarcino's, a 106-year-old ice cream shop and deli that's been in Beth Lagomarcino's family for four generations. I tell Beth how much I admire business owners like her, who persevere despite laws and regulations increasingly stacked against them. Beth is gracious. She's heard about the March and insists on giving me a bag of homemade candy to share with other marchers — a kindness I accept even though it adds weight to my satchel.

Back on the river trail, Grace rings and we chat briefly. She's called often since I left Des Moines, and asks if she can call back tonight when she'll have more time. I say yes, even as I suspect frequent phone calls with someone you've broken up with might be ill advised.

The morning's rain lends a freshness to life along the river. With occasional foot traffic on the trail, I enjoy a pleasant balance of natural beauty and human interest.

One-hundred yards ahead, I see a man and woman walking toward me. She's moving her arms a lot, as if pointing at things. The man's arms hang at his side. When 50 yards away, it appears they're engaged in heated conversation. At 25 yards, I can see that she's angry, her arm movements punctuating sharp words that I can't make out. The man's eyes are downcast, his arms limp in a state of surrender. As the couple passes a few feet away, she says something about custody and visitation while the man continues to look straight ahead, saying nothing.

I find myself wanting to fill in the blanks. Maybe he's a child-support scofflaw who's treated her badly and deserves every bit of her wrath. But maybe she's an angry, bitter person and he's afraid to confront her on the issues that divide them.

The bottom line is I have absolutely no clue what's going on. It's so easy to assume both good and bad things about people you know only from a distance, whether that distance is temporal or spacial. I recall reviewing the profiles of marchers in the months leading up to the start of the March. My assumption about each was along the lines of, "She/he has to be one of the best human beings on the planet to be so concerned about climate change as to walk across America." That, of course, was mostly a reflection of how I felt about my own commitment to undertake this journey.

As we grow closer to others we see their blemishes, flaws, and shortcomings. In a word, we become aware of their humanity — and we're repelled by it because we once thought, wrongly of course, that that person was somehow a step above the usual human cut, just as we're often inclined to believe the same about ourselves. The challenge — my challenge — is to resist the tendency to categorize, demonize, or lionize. The judgmental mind is the unenlightened mind.

Later that day, I pass a young couple on a park bench. They sit together quietly, looking deeply troubled. My brain immediately tries to sort it out. Is she pregnant? Are they about to break up? Did they just have a fight over what color to paint the living room?

The possibilities are endless, and my ignorance abysmal. Speculation is worthless. I can't criticize what I don't understand. I can only accept and love.

I think about my two divorces. After each, I lost friends who made assumptions based on what little they knew, reflected in the mirror of their own experience and bias. After the pain of losing those friends, I vowed never to end a friendship due to conjecture, never to judge others simply on appearance. But that's easier said than done. Cultivating an attitude of love and acceptance requires constant vigilance, frequent self-analysis, and a willingness to confide in friends and perhaps even professionals.

Today, walking alone at a leisurely pace along the banks of America's most powerful and storied river, I confess my flaws even as I accept them. I love myself despite those flaws, and I thank the Great River for its gift of life, for the waters that provide sustenance to all without judgement or bias.

A few of us have home stays tonight. John Jorgensen and I land in a simple home by the river. I sleep well and wake early the next morning, quietly finding my way downstairs to the living room couch. I look out the window at the Iowa shore, visible in the foggy first light of dawn. This is the last I'll see of her for many weeks. My bones ache from the cumulative impact of so many miles. I yearn to go home.

The Great River seems to affirm my desire. I feel it pulling at me, seductively telling me I've walked enough, suffered enough, been lonely long enough. The waters splashing against the muddy riverbank seem to say, "Come home, come home, come home."

Still groggy from sleep, I listen intently. In the voice of the river, I hear Grace. *She's* the one calling me home, begging me to walk no further, asking me to be with her.

But I know that her voice is an illusion, luring me away from duty, inviting me not home, not to love and comfort, but to thin, fleeting moments of gratification, inviting me to the faint, whispered possibility of a life together, a possibility clothed in deceit and disappointment now and as far into the future as I can see.

Grace phoned last night, as she said she would. She called via Skype, and we talked as she bathed in her tub. Her image again rises fresh in my mind, and I feel my senses tingle as they tingled last night. My eyes focus on the far bank of the river. I squint hard, and I'm sure I see her there. Yes, I'm certain of it. My mind fixates on Iowa, drawn in by Grace's sweet, soft voice, her even softer flesh, her rich, brown, hungry, lonely eyes, drawn in by the sight and sound of the bath water lapping against her breasts, much as the warm waters of the Great River now lap against the rocks of the Illinois shore. I fight a desperate urge to run from the house, jump into the river's roiling waters, and flail my way to the far side, to Iowa, to Grace.

Last night, I wanted so badly to be there with Grace, to melt into her song forever. But the invitation is a lie. Yes, Iowa will embrace and welcome me home when I'm done marching. But the embrace of my beloved will be through the screen of a phone or a computer, or with an occasional brief visit to taunt me with the hope of true love, to dangle the dream of domestic tranquility before me only to yank it away as the backdoor clicks shut and the lingering scent of food, sex, and incense is all that remains.

Now I hear the voices of others coming down the stairs and I stir from my dreamlike state. I feel relieved that they'll find me on the couch — not thrashing my way across the river like a drowning man, or broken and battered against the rocks of the Iowa shore. They'll find me awake and ready to march, alone perhaps, but with determination and purpose.

CHAPTER 33

WALKER

> "This made her remember why people take up walking:
> It is because they no longer have anywhere to go."
>
> — CHUCK KLOSTERMAN, ***DOWNTOWN OWL***

In the Sonoran Desert, the population of One Earth Village had fallen to twenty-two marchers. By the time we cross into Illinois, we've grown to nearly fifty. My original dream of 1,000 marchers isn't going to happen. But if Chicago goes well, perhaps we could swell to a hundred for the final stretch. With much on the line in Chicago, marchers are eager to help. Some form an ad hoc committee that meets a couple times in Iowa and is scheduled to meet again tonight.

I enjoy one last glimpse of Iowa as today's 19-mile march leads away from the Mississippi River en route to Erie, Illinois. Jordan Parker, our state coordinator in Chicago, invites us to stop at an ice cream parlor in Erie where she welcomes us to Illinois by picking up our tab. I'm disappointed to learn later that only three marchers wrote to thank her.

Before dinner, I join a dozen other marchers to talk about the Chicago rally. After a brief discussion about speakers, Gavain says, "I'd like to move back to our conversation about disruption."

I'm confused, but it soon becomes clear that "disruption" is Gavain's term for "civil disobedience." At an earlier meeting, he proposed that marchers board a commuter train during rush hour and prevent the train from moving by physically blocking its doors.

"This would send a strong message that all Americans are complicit in our collective inaction against the threat of climate chaos," argues Gavain. Other marchers nod in agreement and share thoughts on how best to implement the action.

When it's my turn to speak, I say, "Honestly, I'm appalled. Let's say you succeed at delaying a train for a half hour, maybe even an hour. You aren't disrupting the life of the oil executive. He's driving to work in his Lexus. You're disrupting a single mother on her way to class at the community college where she's got a big exam. The stodgy instructor doesn't give a damn why she's late and flunks her.

"You're disrupting some old guy who misses a medical appointment he can't reschedule because the free clinic is packed with poor geezers like him. Two weeks later, he dies because the doctor didn't catch a problem that could have been treated, if only the guy had made his appointment."

Other examples of hypothetical victims pop into my head, but I pause and say, "Think about it. Are these the people whose lives you want to disrupt? Do you want to bring new people into our movement, or do you want to alienate all of Chicago?"

Heads that nodded in agreement with Gavain now nod in agreement with me.

But Gavain's not done. He counters, explaining, "This March has been a crucible of transformation and evolution for marchers and many of us are yearning to create an extraordinary moment. We want to define a new type of action that might possibly address this desperate climate situation. It isn't about me wanting self-aggrandizement. It's about recognizing that we all have to rise to the occasion in a new way."

I've learned from my work in politics that when someone says, "It isn't about me," what they mean is, "It's all about me." The discussion goes on for nearly an hour, with arguments on both sides and plenty of meandering excursions that have nothing to do with Chicago or climate change. In the end, only a few marchers support blocking trains.

Crazy talk aside, I understand marchers' desperation. Like Jesus on the road to Jerusalem, some have felt an apocalyptic expectation that, when the March arrives at the White House, the political establishment will simply concede and launch a full-scale mobilization against climate change. Now, with only 20 percent of our journey before us, the reality is sinking in that there will be no grand, transformational finale. This realization, coupled with marchers' lack of organizing experience and the sheer physical and emotional exhaustion built into our lives, has led to what Miriam calls "Marcher Mush Mind."

The next day, I debrief the meeting with Jordan. "I heard about the train-blocking idea a couple days ago," she informs me. "I told the committee it was horrible, but Gavain and a couple others kept pushing. Honestly, I was frustrated you weren't there to back me up."

"Yeah, I've been in my own little world the past few days," I confess. "I'm really sorry. I feel like I've let you down. Let's plan to talk on the phone at least once a day through Chicago."

"That would help," says Jordan, who is now on the verge of tears. "The committee's focus has gone from helping us with the rally to proposing insane stuff to now trying to micro-manage every decision we make."

"I feel really bad about it," I say. "I just don't know how to stuff that ogre back into its cave without igniting a full-scale mutiny, even among marchers who aren't part of the problem. There's a few grounded folks on that committee, but it's dominated by radicals who think crazy thoughts and can stomach interminable meetings."

For now, I hope we've quelled further discussion about blocking commuter trains or any similarly ridiculous ideas.

In Lyndon the next day, the rain holds off just long enough for us to set up our tents. It's my turn to cook, which is zero fun in the rain, especially after a long day's march. Lala's now in charge of buying food. She shows me the vegetables, chicken, and rice I requested, but I've got my doubts about whether it'll be enough for our growing troupe of hungry marchers.

It's not and, as marchers move through the line, I hustle to throw together spaghetti with tomato sauce. Pots of water are boiling and cans of sauce open when Lala barges in, grabs the packages of spaghetti, and barks, "You can't use these. This is tomorrow's dinner."

I bark back, "You'll have plenty of time tomorrow to buy more spaghetti. If you didn't want me using food intended for another meal, you should have bought enough for tonight."

"Marchers aren't going to want to eat spaghetti two nights in a row," she yells at me.

"Hungry people don't give a shit if they eat spaghetti all fucking week," I yell back. "Go away. Let me do my job. Tomorrow you can do *your* job and maybe get it right this time."

Lala storms off as I continue to prepare the spaghetti, mumbling to myself, "You won that battle, Ed — and sure looked like an ass doing it."

The next morning, Lala and I resume our food fight. I'm preparing French toast when she says, "We don't have a lot of eggs. So only use a dozen. Just dip the bread lightly in the egg batter."

I say nothing as I crack three dozen eggs into a large bowl, soak the bread in the batter, and use two loaves more than Lala had allotted. When she finds out, she's furious and screams, "I told you to use only a dozen eggs!"

"Buy another two dozen when you're out shopping for spaghetti," I snap back. "Seriously, are you trying to tank this March by starving people?"

"Money's tight," Lala shoots back.

"Money's always tight, but we have enough for food," I reply. "If you're not willing to buy what we need let someone else have the job."

I'm calmer than last night, but barely. I should have said something positive to ease the tension. I could have reminded Lala of the wonderful meals she made during the first three months of the March, especially that memorable Mediterranean feast on the high desert in Arizona. I could have said how grateful I was in Los Angeles when she, Marie, and Debaura tackled the gargantuan task of outfitting the food truck. I could have reminisced a bit about the wonderful salmon dinner she prepared in New Mexico. But I say nothing and wonder what other marchers think of what must seem like a heated domestic dispute between two very estranged and stubborn lovers.

A MEMOIR FROM THE GREAT MARCH FOR CLIMATE ACTION

After breakfast, the town drunk shows up as we're preparing to march. He teaches me a new fundraising technique. He raises one hand in the air holding a can of beer and gesticulates fiercely with the other, professing his passionate appreciation for what we're doing. He's going to walk five miles with us today and will donate a hundred dollars. I've had plenty of experience with drunks. This one is boisterous but seems harmless. Kathe stays behind to help break down camp, and I advise her simply to ignore the man.

I later learn that the drunk proved good on his word and gave Kathe a hundred-dollar cash donation. Shortly after that, he became belligerent and claimed we were phonies. He demanded his $120 back. Yes, he demanded back $20 more than he "donated"! Kathe obliged and gave him $120 just to get rid of him. I later tease her and suggest we should hire the drunk as our director of fundraising.

I walk with other marchers at first, then veer off on gravel roads to make phone calls. Suddenly, on a particularly quiet backroad with almost no traffic, I hear gunshots. The shots are loud and disturbingly close. I doubt people ever walk this road, and I wonder if the shooter even knows I'm here. In a booming voice, I start singing the first song that pops into my mind: "Mary Had a Little Lamb." I change the lyrics a bit, bellowing, "Merrily we walk along, walk along, walk along. Merrily we walk along, singing silly songs."

The gunshots stop, but I'm shook up and look for a place to sit. I pass two pillars framing a stone wall on either side of a driveway that disappears into a lightly manicured old-growth forest. The noble entrance implies a sprawling estate hidden in the trees, a place of wealth, history, and privilege. I hop onto the wall and sit under the shade of a stately oak. The shade and the cool stones calm my rattled nerves. I know I'm on private property, but I don't see a house or the owner and decide that this is an excellent place for lunch.

As the juice of a pear runs down my chin, a vehicle approaches quickly from beyond the trees. "Busted," I say to myself. "No point in moving now. Just sit tight and hope for the best."

A large van rolls through the pillars as the driver pulls up beside me. My mind races. "Is this the guy who was firing the gun? Is he going to call the police because I'm trespassing? Worse yet, am I about to get a taste of vigilante justice?"

The man slowly opens his window. He's white, middle-age, and heavy-set. In an unexpectedly jovial voice, he smiles and says, "Did they just drop you off and leave?"

I laugh, relieved, and tell him I'm with a group walking across the country for climate action and we hope to be in Chicago in a few days.

"Well, I'd give you a ride," the man says smiling, "but I'm not heading that way today."

"Thanks, I appreciate that," I reply. "But I'm committed to walking."

There's a momentary pause, and I'm curious so I ask, "If you're not heading to Chicago, where *are* you going?"

"Oh, I'm just picking up my mail," the man says, as he does a U-turn in front of me, swings by his mailbox, and collects his mail without ever getting out of the van. "Have a great walk," he says as he drives back to the house.

"Nice enough guy," I say to myself as I hear the van stop not more than 75 yards away, just beyond the trees at what I imagine to be a spacious and luxurious mansion. I'm guessing that the distance the man walks getting around the interior of his grand home every day is far greater than the distance he just drove to retrieve his mail.

It's a beautiful day. Temperatures are in the upper 70s with sunshine and a light breeze. I stare up the driveway and shake my head.

For a couple weeks the March had enjoyed a spell of harmony and sanity. But as we approach Chicago, One Earth Village spirals to stupidity and beyond:

- Jordan drops $280 of her own money to treat marchers to a drink at a bar next to one of our campsites. Nearly every marcher enjoys a drink, but only three thank her.
- A restaurant manager goes to great trouble to arrange for the Climate Justice Gypsy Band to perform, only to

have two band members cancel the gig at the last minute without consulting the rest of us.

- Jordan has worked hard to line up home stays for marchers outside of Chicago. With only one day's notice and no discussion with me, Kim and Liz tell her to cancel the home stays.
- On short notice, marchers want staff to call in to a meeting with detailed, written activity reports. Staff drop other work to accommodate the request only to have the marchers who asked for the reports cancel the meeting 15 minutes before it was to start — then uncancel it 10 minutes later. "Third-world dictators have a better grasp of democracy than some of you," I mutter within earshot of the offending marchers, promising never again to waste my time on such requests or meetings.

The meddling and micromanaging of Jordan's work gets worse by the day. To accommodate our sponsors in Chicago, she has to allow six speakers at the rally. Some marchers want only four. When Gavain learns of Jordan's decision he writes an angry message: "I am not here to promote some second-tier progressive organization that is inconsequential to world events." I presume Gavain is referring to the Sierra Club, Citizens' Climate Lobby, the Chicago Botanical Center, the Field Museum or one of the other ten renowned "second-tier" partners Jordan and others have lined up.

Piling on, Gavain's partner, Dana McGuire — a roller derby competitor who goes by the name "Greenhouse Gashes" — writes to Jordan, "We are all displeased with you washing aside our considerations about the number of speakers, as well as your apparent general disrespect of the marchers."

It's remarkable that some marchers have developed such a strong sense of privilege. Staff have worked so hard, and people in general have been so consistently kind to us, that many marchers now take it for granted. Like members of a dysfunctional royal family, some simply *expect* to be admired and coddled.

The internal health of the March has decayed irreparably this week.

Could I have prevented it if, back in Los Angeles, I had chosen not to march or host my talk show and instead focused strictly on directing the March? I don't know. But admittedly, attempting to do everything I've taken on is more than one person can juggle. I've kept the March cobbled together since California. But as the seeds of organizational decay sprouted and spread, the March's structure inevitably crumbled to the point that it couldn't be fixed.

In Des Moines and again on the Mississippi River, I had wanted to quit. Coming into Chicago, I want to quit again. This time the sense is more compelling than ever. As I walk in silence I arrive at a comfortable compromise.

I'll quit the March, at least in my mind. I'll stop being the director, stop raising money, stop caring if marchers' actions are smart or stupid. I won't hire coordinators for the remaining states. We'll rely more heavily on donations that come in from town to town. I'll make sure there's enough money to buy food, pay current staff, and meet existing financial commitments, but no more.

"That's it." I tell myself with a profound sense of relief. "I'm no longer a marcher. From now on, I'm simply a walker out for a 700-mile stroll. If all goes well, I'll find myself in Washington, DC, on November 1."

A MEMOIR FROM THE GREAT MARCH FOR CLIMATE ACTION

CHAPTER 34

SACRIFICE ZONE

"Race and class are extremely reliable indicators as to where one might find the good stuff, like parks and trees, and where one might find the bad stuff, like power plants and waste facilities."

— MAJORA CARTER

Only a couple dozen local supporters walk with us into Chicago. In the heart of downtown, some marchers veer off on what feels like a detour. The rest of us follow. As we pass the Chicago Board of Trade, marchers string yellow hazard tape across the entrance and stage a die-in. I stand quietly nearby. The March has indeed morphed into a beast over which I have little control. Local participants are confused. Some mutter, "What's going on? I don't get it. We weren't told about this." A local supporter asks me to explain the message. "There's no message," I say, "only desperation."

We arrive at Daley Plaza where a small crowd joins us for the rally. Gavain is one of four marchers to speak. With great drama, he says, *"We. Are. Dying."* What follows is the most depressing rally speech I've ever heard. Kelsey also speaks, her remarks peppered with abundant references to extinction and hopelessness, but failing to match Gavain's command of fatalism and morbidity. I'm the last speaker and try to end the rally on a positive note, citing effective actions across the country and thanking our local volunteers and partnering organizations.

I sleep a lot during our day off, and feel refreshed as we set out with six new marchers on September 8. Local organizers warn Sarah that our route takes us through neighborhoods with a recent history of gang violence. She begs marchers to stick together. Everyone complies.

Tenting is not an option tonight. We're fortunate to land floor space at a Methodist church where the congregation generously serves us dinner. In another display of rudeness and privilege, half the marchers arrive nearly an hour late, keeping everyone waiting. I suggest to a few that they might want to apologize to our hosts. One responds, "Yeah, we'll apologize — for trying to save the planet."

"It's not about saving the planet," I shoot back. "It's about basic respect and courtesy." I apologize to the pastor as the late arrivals line up for dinner.

There are no showers at the church, but we're able to use the bathrooms. By now, I'm a pro at vanquishing the day's sweat and grime without the benefit of a tub or shower. I'm washing my feet in the sink when a new marcher walks in. He's a big guy with long, dark hair and a rugged demeanor. "Hey, I'm Fernando." He looks at my feet in the sink as he fumbles for additional words. "Well, that's kinda weird," he says, laughing a bit uncomfortably. "Is that how you always do it?"

"No," I respond curtly. I'm tired and not in the mood for criticism. "I'm spoiling myself tonight. Usually, it's a quart of cold water and a handkerchief while I'm squatting behind a rock with mosquitoes gnawing on my ass. Welcome to the Great March for Climate Action."

Fernando laughs for real this time, "Well, I guess I better learn the technique, 'cause I'm with you all the way to DC."

I smile now, happy to hear that a new marcher with a sense of humor plans to go the distance. Fernando talks as I scrub. Most marchers are from wealthy or at least financially comfortable backgrounds. Fernando's cut from a different cloth. He's an ironworker from Chicago, half Mexican and half Potawatomi. He can't work anymore and shows me his badly mangled left hand. Four years ago, it was crushed by a huge piece of machinery. Ever since, he's been in a legal battle with Imperial Steel.

"On construction sites, ironworkers were the top dogs," he tells me. "We did some of the most dangerous work. And ya know, there was pride in that, and ego. Ironworkers get hurt all the time. A buddy of mine once fell 40 feet onto concrete and broke his tailbone and pelvis. Me, I only had minor injuries for twenty years — until I totally fucked up my hand."

I look at Fernando's hand more closely. It's hard to imagine what kind of pain that injury involved. "Yeah, the hand thing sure messed up my life," he tells me. "I had anger issues before then, but sometimes they're worse nowadays. At least I gave up drinking. Mostly."

"Well, we need you on this March," I say. "I'm glad you're here. Make sure you get a good night's sleep, 'cause we've got a long haul tomorrow through hell on Earth."

"Yeah, I've seen hell on Earth," Fernando laughs. "In fact, the Devil used to sign my paycheck."

The 19-mile march to Gary, Indiana takes us through East Chicago and Whiting, past the most disturbing industrial carnage most of us have ever witnessed. We pass miles of smoke stacks and flaming towers — garish monuments exposing the lie behind modernity's sanitized facade. Of all the ugliness we've seen on the March this is the worst. It's also the most relevant to our purpose and mission. Only Fernando is familiar with the area and doesn't seem shocked. Many marchers are overwhelmed. Some cry.

The break truck waits for us as we pass the main gate of the BP refinery. Several marchers grab our large Climate March banner and unfurl it across the entrance as others rush over to help hold it up. Two large trucks approach and security officers order us to let the vehicles through. Most marchers move out of the way, but Sean, Mack, and two others sit in front of the gate, blocking traffic for 15-20 minutes until police arrive and threaten to arrest them if they don't move. Eventually they do.

Unlike the die-in in Chicago, this spontaneous action makes sense to me. With the exception of Fernando, how could any of us have known how powerfully these refineries would impact us as we walked by, close enough to smell the toxic fumes and hear the flames shooting into the air?

Later, we stop at the home of Thomas Frank, a nearby resident who has helped organize opposition to BP's expansion plans. "You just walked through a gated industrial community," Thomas says. He tells us that the sprawling complex of oil refineries and steel mills stretches twenty miles along Lake Michigan by neighborhoods that are ninety percent Black and Hispanic.

"The water here is badly contaminated with lead and other heavy metals," he says. "The air quality in Lake County ranks as one of the worst in the nation, and we've got one of the highest infant mortality rates, too."

Thomas goes on to explain that, earlier this year, BP completed a $4.5 billion expansion of its refinery. "Now they want to tear down 100 homes in a national historic district and displace 400 people. They just don't care — and they've bought off the politicians, who don't care either."

Walking away from the main gate of the refinery, I see a patch of ground that at one time might have been a small parking lot, last used perhaps twenty years ago. Life has sprouted between the cracks of the pavement. Sunflowers, goldenrod, a dozen or more other species, and a soft, wavy grass I don't recognize are gradually reclaiming this tortured sliver of Earth. Through the patient power of life and rebirth, these unassuming plants are toppling this kingdom of concrete, undermining an empire that only a few years ago seemed unassailable.

I pick one of the grass's seed heads and run its soft, furry surface through my fingers. I'm nurtured by its touch and encouraged by the simple fact that this wand of grass lives. It remains my companion for the next few miles. I shift it from one hand to the other, every so often scattering a few of its seeds. Occasionally, I carry it between my teeth, biting down gently to taste the grass's sweet juice. I feel sadness for those who live next to these hideous refineries. I offer a silent prayer that, despite the greed and madness that created this sacrifice zone, life here will survive and, in time, thrive.

We leave Gary, Indiana early the next morning on a bleak, rainy day, walking along busy streets during the morning rush hour. Suddenly, Mack and Sean scurry past me on all fours. My first reaction is surprise at seeing Mack without crutches. Then I notice duct tape covering their mouths. Forgetting my promise not to care how other marchers look and act, my incredulity gets the best of me. I say,

"Please guys, what the hell are you doing? If you're trying to make a statement, I have no idea what it is and neither do all these people driving to work."

Mack and Sean say nothing as they slink past like a pair of half-human creatures. I later learn they crawled for over a mile, at which point Mack resumed walking with crutches. Their apparent intent was to demonstrate solidarity with the innocent, silent beings that climate change threatens. I share their distress, but I'm certain no passerby had any idea what they were trying to say. Their action strikes me as profoundly self-indulgent and makes all of us look silly.

Between East Chicago, Illinois and South Bend, Indiana, we have four long days: 19, 22, 24, and 17 miles. This might explain the resurgence of plantar fasciitis. I feel fortunate that, east of Gary, the roads are mostly beautiful and quiet, often with a grassy strip along the side where I walk to ease the discomfort in my feet.

Life below my knees is tough for another reason. Insects have been feasting generously on my feet and ankles. I suspect chiggers, mosquitoes, and fleas. Most are as elusive as ninja warriors, and I rarely catch the offending mini-monsters. I have bites on bites — more bites than I've ever had, more even than when I lived in northern regions where swarms of mosquitoes, black flies, and no-see-ums were common. The bites are causing my feet to swell. Between that and the plantar fasciitis, walking is painful and my pace slow.

I'm relieved to have a day off in South Bend. It's again my turn to cook supper and breakfast. After my most recent spat with Lala, she's given up nagging me. Instead, she's cold and standoffish, and I revel in how our relationship has improved.

This afternoon, without the time crunch of a day's march, I unleash my culinary muse in the fully equipped kitchen our hosts let us use. I don't want marchers to go hungry, so I prepare way too much food, knowing that leftovers will find their way into marchers' lunch boxes or tomorrow's dinner. Perhaps, too, cooking so much food is how my petulant subconscious hopes to "win" another victory over Lala, this time without the undignified display of a ranting string of cuss words.

The next morning, I'm up shortly after 4:00 to make breakfast — a feast of fried potatoes, pancakes, and made-to-order eggs. Marchers aren't used to having a choice and are genuinely surprised when I ask how they want their eggs cooked. Breakfast takes longer than usual, but most appreciate the customized service.

Most people who walk, run, or bike hard know that spitting is an essential part of physical exertion. I try to be subtle when I spit, try to do it when no one's looking, aim for grass or dirt. Walking today, I spit and fail to notice a woman sitting in her car by the curb. She eyes me sternly and says..."That's disgusting."

"Sorry," I say, wondering if she detects the insincerity in my voice. "Nice car," I offer.

"Um, thank you," she responds, thrown off, her phrasing more of a question than an answer.

"Your car reminds me of *my* vehicle," I continue.

"Really?" she asks. "You have a Nissan Xterra?"

"No, no. Your car reminds me of *me*. My shoes are the tires, my socks the shocks. My eyes are the windshield, my eyelids the wipers. My stomach is the fuel tank, this walking stick is my four-wheel-drive, and my mouth —" I pause for effect. "My mouth is the tailpipe," I nod with a manufactured sense of deep satisfaction.

She stares at me, squinting a bit, a little confused, a little amused, and says, "And your point is?"

"Tailpipes!" I say loudly as I raise my arms and walking stick. "That's the real difference between your vehicle and mine. Your tailpipe spews toxic fumes. Mine spews saliva — and that's 99 percent water."

I pause briefly, then add, "My emissions may look disgusting to you, but they're harmless. Yours on the other hand *are destroying the planet!*" I say these last words with great emphasis, and for a hot moment feel I'm channeling Gavain.

The woman mutters something I could have sworn was a muffled expletive, shakes her head and drives off, her tires squealing that petulant, pouting squeal tires make when they're pissed off and want to run over your face but, out of fear of legal consequences, take it out on the pavement instead. The moment is priceless — yes, priceless, and childish. I've just squandered an opportunity to have an intelligent conversation about climate change, but today I don't care.

Later, I pass through a small village, a cluster of a dozen or so homes. Some are in good repair, others poorly kept. All feel cozy, even the shabbiest — a small, white box with exposed boards thirsting for paint. In a cracked window, a shade yellowed from years of sunlight hangs off kilter. Weed trees that years ago sprouted an inch or two from the foundation now push threateningly against the basement walls. A tarnished satellite dish perches awkwardly on the roof, like a facial wart grown discolored and more prominent with age.

This is not a place most people would want to live. Yet I imagine the inhabitants find comfort and security here. It's their *home*, after all. I slow my pace, and though I see no one, I'm certain many eyes watch me, peering from behind the corners of shades and through holes in curtains. I return the hidden townsfolk's cautionary stares and curiosity with a twinge of jealousy, glancing at their homes, drawn in by imagined warmth, tea, stability, tranquility, and the presence of loved ones.

As I pass the last house in the village, my perception shifts and I see it all differently. These aren't nests of peace and contentment. They're traps! They're cages sealed with a mortgage and fear of the unknown, cells for lonely people afflicted with discontent that, over time, will melt into resignation. Wouldn't the prisoners trapped here love to trade places with me and taste the freedom of walking along an unfamiliar country road?

I feel a distinct moment of clarity, as if the root cause of human unhappiness has been revealed. Stability and comfort are empty without adventure and purpose. We need both if we are to be truly happy. That's it, isn't it? Our spirits will remain troubled as long as we believe we must have just one or the other. Deep happiness comes with recognizing this dichotomy and choosing to live fully in the present, whether one's present is settled or unsettled.

I walked into this village sore and tired, craving the comforts of home. I quicken my pace as I pass the last house and move out of sight from the hidden, staring eyes that I imagine watch me and wonder, perhaps even with a touch of envy, who I am, where I'm going, and why I walk. I find peace not knowing what lies around the next curve in the road.

MARCHER, WALKER, PILGRIM:

Bob Cook with lunch generously provided by the residents of New Buffalo Center, Arroyo Hondo, New Mexico.

Ed and Steve in silhouette.

Ed in front of the museum at Fort Garland.

MARCHER, WALKER, PILGRIM:

Ed pretending to be roadkill.

The Climate Justice Gypsy Band performs at our rally in Colorado Springs: Kat, Mack, Luke, John J, Lala, Sean, Izzy, Ed, and Jimmy. (Band members not pictured: Faith and Berenice.)

A MEMOIR FROM THE GREAT MARCH FOR CLIMATE ACTION

Storm clouds menace our camp in eastern Colorado.

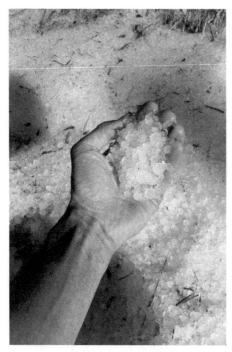

On two occasions, hail fell so thick we could gather handfuls of it the next day.

One huge hailstorm decimated crops along our route.

In eastern Colorado, Steve leaves the March and sets out for New York on his own, pushing his gear in a cart.

We march in the Independence Day parade in Culbertson, Nebraska.

Berenice pushing her gear in a stroller.

Besides leading the Climate Justice Gypsy Band, Izzy was known for his passion for large bags of chips.

March massage conga line: Chris Ververis, Jimmy, Mack, Anna Wishart's dog, and Liz.

Marchers celebrate as we cross the path of the Keystone pipeline near York, Nebraska.

Crossing the Missouri River into Iowa from Omaha, Nebraska.

David Osterberg.

A MEMOIR FROM THE GREAT MARCH FOR CLIMATE ACTION

Left to right, back row: Ed, Dana McGuire, Izzy, Gavain U'Pritchard, Ben, Kelsey Erickson, Zach Hough Solomon. Middle row: Jeffrey, Lala, Kelsey Juliana, Berenice, Tilly (Gavain and Dana's daughter). Front row: David Osterberg, Chris, Faith, Lee, Mack.

The March's mobile wind turbine in front of Gavain's vegetable-oil-powered truck.

MARCHER, WALKER, PILGRIM:

Locals mingle with marchers at Ritual Cafe before the two-mile parade to the Iowa State Capitol.

Jeffrey.

Ben and his dog Diggy with Ed in front of the Iowa State Capitol.

Davenport Mayor Bill Gluba leading marchers across the Mississippi River.

The Climate March banner hangs over our camp on the Illinois side of the Mississippi River.

Marchers stage a die-in in Chicago, Illinois. (Photo by Bob Simpson)

Faith and Paul Moeller in front of the BP refinery, Whiting, Indiana.

MARCHER, WALKER, PILGRIM:

Marchers block the entrance to the BP refinery with the Climate March banner.

Kathe, John A, and Miriam fend off the worst attack of mosquitos outside of Chicago.

Faith sitting beside her mangled cart outside of Toledo, Ohio.

Fernando Cazares and Ed.

Miriam and Mack.

Ed and Sean.

Jeffrey and Miriam.

Rally against fracking in Butler, Pennsylvania.

Doug Cooley next to the March's solar collector.

A MEMOIR FROM THE GREAT MARCH FOR CLIMATE ACTION

Fernando horsing around.

Kelsey Erickson.

The C&O Canal Trail in Maryland— Ed's path for three days during the last week of the March.

A MEMOIR FROM THE GREAT MARCH FOR CLIMATE ACTION

Halloween decoration on White's Ferry near Leesburg, Virginia.

The final mile of the March in Washington, DC.

At Lafayette Square across from the White House: Dana McGuire, Bruce Nayowith, Anita Payne, other marchers and local supporters hold a string of postcards filled out by people we met along the way. The postcards address the urgency of climate action.

Marchers and local supporters pose for a photo in front of the White House after the rally.

MARCHER, WALKER, PILGRIM:

CHAPTER 35

LONGEST MEETS LARGEST

"There comes . . . a longing never to travel again except on foot."

— WENDELL BERRY, *REMEMBERING*

Five of the ten largest Amish communities in North America are in Indiana. Our route takes us through the most concentrated of these in Elkhart and Grange counties. For two days we're immersed in an agrarian paradise of small, tidy, sustainable farms. We travel slower than usual, instinctively adjusting to the more relaxed pace of Amish life.

A few of us stop to watch Amish youngsters play baseball. On another occasion, a man walking his bike pauses to answer questions about Amish farming practices. In Middlebury, I strike up a conversation with two women who ask where we're going. I tell them, "Washington, DC!"

"That's a long walk," one of them replies matter-of-factly. "Me and my husband are taking the train there next month for our vacation." She excitedly describes some of the sights they plan to see.

"Well, if you stay long enough, we'll meet up with you at the White House," I suggest.

She laughs, and hands me three bananas. "Here, you'll need these to keep you going."

Among all American demographics, with the possible exception of cave-dwelling hermits, the Amish surely have the lowest carbon footprint. As I admire the many front yards adorned with laundry drying in the late-summer sun, I'm reminded that if all Americans traded their electric or gas dryers for a clothesline and drying rack, we'd save 6.6% of the country's total residential output of carbon dioxide.[1]

In more "civilized" places, hanging laundry is often frowned upon as tacky and low-class. The anti-outdoor-laundry bias is particularly evident among America's 300,000 homeowners' associations, many of which ban clotheslines outright.[2]

Outside of Shipshewana I run into Kelsey as she's resting from pulling the rickshaw. It's loaded heavy with marchers' gear, and I offer to give her a break for the last mile. This is my first time pulling the rickshaw and I'm surprised at how easily it rolls.

Kelsey and I talk a bit as we walk into town. We've gotten along well since she rejoined the March. She knows I've consistently defended those who use the truck to transport their gear, but she also appreciates that I've praised her and other marchers willing and able to haul their own.

We see an Amish buggy approach with a full complement of passengers. As the buggy passes, the horse sees me and is badly spooked — presumably because she's never encountered a buggy pulled by a human. The horse begins to bolt and it looks like the buggy might flip over. The skilled driver manages to bring the horse under control. It's a frightening moment. The Amish man yells at us to move to the other side of the road. We don't since that would cause a whole different set of problems. But the next time we see a buggy coming, Kelsey and I stop and wait until it passes.

Arriving at our camp at the Shipshewana Auction and Flea Market, Sarah laughs to see me pulling the rickshaw. "Hey guys," she says, "you should visit the auction tomorrow and buy a horse or two, maybe even a donkey, to pull more carts."

"Great idea," I reply. "When we get to DC, we can donate them to the Obamas to provide fertilizer for the White House garden."

Tonight, I pitch my tent too close to John Jorgensen's. John's a powerful snorer, so I preemptively dig out my earplugs as I crawl into my sleeping bag. As I'm drifting off to sleep I hear what sounds like Catholic monks chanting in the distance. Startled, I sit up and

---

[1] https://www.thespruce.com/reasons-to-line-dry-laundry-2145997

[2] http://www.nytimes.com/2007/04/12/garden/12clothesline.html

remove the earplugs. Monks? At an Amish auction complex? It takes a few seconds for me to realize that the sound isn't coming from a chapel full of monks but from a barn full of cows. I laugh as I reinsert the earplugs, thinking that, in terms of white noise, I've got a better shot at a good night's sleep with happy monks than unhappy cows and John's snoring.

Indiana is the first state we've navigated without a paid coordinator. Many people have stepped forward to help but none more than Clifford Peterson. He and his wife, Lisa, have worked tirelessly with Sarah to line up lodging, potlucks, speaking opportunities, and more.

As we cross into Ohio, Clifford sends marchers a farewell message. "Whatever shape the end of the March turns out to be, you all have profoundly influenced our lives and Indiana vastly for the better. Many of the effects of the March are ones you will never know. History will remember the March as the advance guard of a great turning point in the course of events. The current schizophrenia of the group is perhaps inevitable, however painful it may be to you at this moment and however sore your feet must be."

We're encouraged by Clifford's words. I appreciate his candor about our "schizophrenia" — a condition that again becomes evident as we prepare to take a four-day break at the Indiana-Ohio border to travel to New York City for the People's Climate March.

Hundreds of thousands are expected at what is billed as the largest climate march in US history. When first announced last spring, marchers discussed how we could participate yet still finish our own march. Some wanted to raise funds to rent a bus. I was open to that, and thought organizers in New York would welcome the chance to collaborate around the catchy "longest meets largest" message. But consistent with my experience over the years, the National Non-profit Industrial Complex, as I've come to call it, paid lip service to our repeated enquiries and, in the end, completely ignored us.

Two weeks ago in Illinois, at the height of tension within the March, I contemplated two sleepless nights on a bus and decided not to go to New York. The trip would leave me exhausted to the point that

I might not be able to resume walking when we returned.

But lately I've been under a lot of pressure to attend from some marchers and the documentary crew. I settle on what seems like a reasonable plan. I'll borrow a car from Bryant Marcontel, one of the crew's videographers, drive to nearby Detroit, and fly to New York. It's an expense I can barely afford and a compromise in terms of my carbon footprint, but my primary concern is finishing the March.

Predictably, this decision doesn't settle well with several marchers. Sean, who broke her fast of silence in Colorado, is very upset. "I used to respect you," she tells me, tears welling up in her eyes. "I used to regard you as a mentor, and at the start of the March, you told me I was like a daughter to you."

"You *are* like a daughter to me," I assure her. Besides training together in Des Moines before the start of the March, Sean and I also shopped for shoes, played music, and shared meals. "Please try to see the big picture, Sean," I say. "We all make compromises. This march in New York is important, and we're all compromising to get there. I'm taking a carbon-spewing plane while you're on a carbon-spewing bus. Earlier this year, I took a train to the start of the March while you flew. No one faults you for that. It's what you had to do."

But logic is not what Sean needs. She's sobbing now, and says, "But I feel so hopeless and helpless. All we're doing is marching, every day marching, while the planet is falling apart. Two-hundred species go extinct *every day!* And marching isn't stopping it!"

Sean is frantic now, and I don't know how to reassure her. I hug her tightly and say, "Remember, every little thing we do is important. Whether you're marching, protesting, or campaigning for office, you're making a difference. And if you and I have to compromise in the process, let's not be too hard on each other, ok?"

"I'll try," says Sean. "But it just seems like the world is about to end and we need to do something extreme to wake people up."

I understand Sean's frustration. Perhaps I have more patience and see the long haul simply because I'm older. But Sean has good reason to be anxious. Most climate models predict drastic if not disastrous changes in the near future.. It's a fair question: "What will it take to wake people up?" I don't have the answer, but I hope Sean doesn't do something so extreme that it lands her a long prison sentence.

Clearly, Gavain's influence on Sean and other young marchers is increasing. This could lead to crazier ideas than blocking commuter trains. I would feel beyond horrible if Sean or another marcher did something that ruined their lives. Even though I just want to walk and have tried to distance myself from the March's "schizophrenia," *I'm* the one who's unleashed this beast. *My* name is on the March's legal documents. I've backed away from some of the duties of director. But if something goes wrong, *I'll* be held accountable. No matter how badly I'd like to wash my hands of the March and just walk, the responsibility to preserve sanity and keep everyone safe ultimately rests with me.

While waiting at the airport in Detroit, I hear from Steve. I had my doubts he could manage 25-30 miles a day on his own and make it to New York by September 21. Incredibly, he does and his family joins him as he crosses the George Washington Bridge into Manhattan — one day before the People's Climate March. It's an amazing accomplishment.

The People's Climate March organizers assign us a spot toward the back of the March, sandwiched between vegans and communists. I'm excited that Shira's here and that we'll get to walk together again, even if only for three miles.

We wait for four hours to start walking, giving me ample time to visit with marchers and some of our neighbors. One young communist gives me a solid ten-minute lecture on how capitalism is responsible for climate change. I point out that communists have been just as successful as capitalists at environmental destruction. "The USSR killed the Aral Sea, literally wiping it off the map," I point out.

"Yeah, well look at what capitalism did to Lake Erie and the Gulf of Mexico, where there's now a dead zone the size of Massachusetts," he responds.

"Exactly," I say. "Pick your economic poison. The problem isn't capitalism. It's industrialization without checks and balances."

When I tire of fighting communists, I turn my curiosity to vegans. Their signs and chants evangelize the message that climate change will be solved only when everyone stops eating meat. Communists are dogmatic. Vegans, even more so. I opt out of conversation and amble off to a nearby deli for a Reuben sandwich.

Once we're moving, the documentary crew swings into action, filming a half-dozen marchers. Shira and I walk together with two cameras and a boom mic in front of us. Some people assume I'm famous. One woman comes over and says, "You must be somebody important." Shira smiles and says, "We're all important," as she gives the woman a big hug.

Next in this faux-celebrity spectacle a photographer sporting an impressive camera lens gets in my face, between me and the documentary crew at times, furiously clicking away. After a bit he says, "So, I hear you're the founder of this march."

"No, this is the *largest* climate march ever," I correct him. "I'm the founder of the *longest* climate march ever, the Great March for Climate Action." This revelation is met with noticeable disappointment. The photographer has never heard of our march and quickly scurries away as Shira and I enjoy another good laugh.

After the March I head off on my own, wanting to get a good night's sleep before flying back to Detroit. I feel oppressed by the sheer volume of steel and concrete that is Manhattan. Block after block, nothing but steel and concrete, with barely a plant or tree in sight.

Suddenly, floating in front of me is a milkweed seed. Where did it come from? Is there a milkweed plant nearby, perhaps even a whole colony of milkweed? Is someone planting milkweed to attract monarch butterflies?

I follow the seed, sometimes glancing behind me, upwind, hoping to catch a glimpse of some patch of green that might suggest the seed's point of origin. I keep pace with it, smiling, then laughing, not caring that people might think I'm a crazy man. I chase the seed down the street. It slips around a corner, then slides under a magazine rack, then flirts with diving into a subway entrance. I'm ready to pursue it underground when it veers off across Broadway, across a raging stream of cars and cabs. All I can do is watch the seed drift deeper and deeper into a land of financial and material make-believe. I wish I could assist its journey. I feel slightly guilty that I didn't catch it when I had the chance, catch it and plant it somewhere safe and protected.

But that would have interfered with destiny. Perhaps this seed will find its way to a rare patch of fertile Manhattan ground and sprout new life next spring. Perhaps that plant will go on to breed hundreds more plants that spawn generations of milkweed, blessing the world with beauty and feeding thousands, perhaps millions, of monarchs. Whatever happens, like the plants reclaiming the parking lot at the BP refinery south of Chicago, this small, floating, magical seed gives me hope.

"Life conquers despair," I say to myself, as I think of Sean and wish she had been here to share this moment.

A MEMOIR FROM THE GREAT MARCH FOR CLIMATE ACTION

CHAPTER 36

COLLISION

"Four times I was honked at for having the temerity to proceed through town without the benefit of metal."

— BILL BRYSON, *A WALK IN THE WOODS*

After flying from New York to Detroit, I drive back to our camp in Montpelier, Ohio. Thanks to the four-day break I feel refreshed and ready to walk. I arrive just as marchers exit the bus. They've ridden twelve hours through the night. Some look painfully exhausted, dragging themselves off the bus like knapsack- and blanket-toting zombies. I'm crushed by a sudden wave of guilt, like the kid who just scarfed a huge wedge of chocolate cake while his friends choked down tasteless blobs of cheap green jello pock-marked with suspended chunks of processed fruit.

To make up for lost time, we march an average of 18.5 miles each of the next three days. Some marchers catch rides for part or all of a day. Most march despite their fatigue. Six of us — Steve, Jeffrey, Miriam, Lee, Mack, and I — have walked every step of the way. Steve finished his march in New York City. Jeffrey appears indefatigable, despite losing thirty-three pounds and becoming a vegan. Miriam looks so worn down I fear she could simply collapse. Lee and Mack remain in New York, and it's unclear if they'll return to finish the March.

Much of our route through Ohio is on trails. I welcome the traffic-free travel and enjoy frequent snacks of wild food. There are plums, apples, grapes, greenbrier, sheep sorrel, and more. Sean is eager to learn about edible wild plants. I share what I know with her as we walk together. Once, we find three giant puffballs, and marchers pulling the rickshaw haul them to camp for supper.

I'm encouraged that my rapport with Sean, Kelsey, and many of the younger marchers has improved. But there are new conflicts. A growing divide has formed between those of us focused on marching and education versus those who want the March to evolve into a roving squad of nonviolent resisters.

Jimmy is increasingly of this mind. In an email to most marchers (but not to me) he writes, "We have returned from the spectacle that was the People's Climate March in New York City and we have real work to do."

Jimmy calls the People's Climate March a "largely symbolic showing [of people who] simply went back to the comforts of their homes." He downplays the importance of marching, saying we should be about more than "narrowly focused footsteps in a vast, privilege-tainted ecological justice movement." He, Lala, Gavain, and others talk about building a movement that is "leaderless," which is code for "different leaders." When Sarah shares Jimmy's message I smile, shake my head, and say, "Let the coup begin."

"We've got a power struggle going on, Ed" says Sarah. "But that's *your* problem, not mine. I've got my hands full with charting the route, finding campsites, moving vehicles, yada, yada, yada."

"You sure do," I say. Without paid state coordinators, Sarah's workload has increased, which is hard to imagine. She's on task constantly and getting only four or five hours of sleep a night.

"If things get real goofy, like they almost did in Chicago, we'll find a way to keep the March from unraveling," I reassure her, though I have no clear plan as to how that would happen.

After three long days, September 26 is an easy 10-mile trek to Toledo, followed by a day off. Lee returns from New York and meets us in Toledo. Mack insists on walking every step of what he missed. He finds a ride back to Montpelier and, walking on crutches and without a support vehicle, cranks out back-to-back thirty-mile days to catch up with us in Toledo. I'm in awe at Mack, and wonder if the guy is even human.

Toledo is the first large city where we don't have a rally — a clear casualty of not hiring a paid state coordinator. No rallies are planned in Cleveland or Pittsburgh either. It's remarkable to me that the anti-staff element of the March doesn't recognize the connection.

Our host in Toledo is Good Shepherd Catholic Church, located in a neighborhood where increased gang violence is a problem.[1] We're asked not to walk alone and are forbidden to camp. Instead, the Church lets us have the entire third floor of a former convent to sleep in.

Some home stays are available, though most marchers opt to sleep at the convent. I stay with Valerie Crow, one of Toledo's most outspoken Native voices on climate. Valerie's home emanates peace and calm. As much as I'd simply like to rest on our day off, I can't escape a heavy workload. Valerie drops me downtown where Sarah and I meet to work at a restaurant.

After a hearty breakfast the following morning, Valerie drives me to the convent. We arrive as Sarah is going over the route with marchers. Valerie introduces herself, saying, "I'm so grateful for what you are doing. I wish I could join you, but I can't. I've just come to thank you and drop off your leader."

Lala goes ballistic. She yells, "He's not our leader!" A few minutes later, Lala again reminds everyone assembled, "Ed is *not* the March's leader. We're a leaderless group."

I'm annoyed at Lala for making Valerie feel uncomfortable, but I say nothing. As marchers exit the parking lot, I remain behind for a few minutes to thank our other hosts. Lala also remains behind because she doesn't plan to walk today. As I head out, I turn and say with evident drama and a wave of my hand, "Let's go Lala, follow me!"

The look on Lala's face is priceless as she yells, "Stop it! You know that makes me crazy."

Today's march to Elmore is 17 miles, some of it along Highway 51. We spread out as we leave Toledo. I walk by myself in sight of a few other marchers. I see Faith about a quarter mile ahead of me pushing a baby stroller loaded with gear. I lose sight of her for a moment, and

---

[1] http://www.toledoblade.com/Police-Fire/2013/04/28/Gangs-exact-bloody-toll-on-Toledo.html

when I walk a bit farther I see her sitting on the side of the highway. I quicken my pace. As I come closer, I spot Faith's stroller upended and crushed on the side of the road. She's been hit!

I rush ahead. Faith is shaking and crying, but miraculously she's uninjured. A driver ran into the stroller, and the force of the impact knocked Faith through the air. She fortunately landed on grass. The stroller took the full brunt of the impact and quite possibly saved her life. The driver pulled over and a crowd has gathered. People gasp when they see the stroller, assuming a baby is trapped inside the mangled frame of metal and plastic.

More marchers arrive. They console Faith and assure the concerned bystanders that Faith is ok and that the stroller was transporting gear, not a baby. I speak with the driver of the vehicle. He's an older man who at first claims to have fallen asleep. His story evolves quickly and it's unclear what caused him to hit the stroller. Since Faith is uninjured, I suggest that he simply pay to replace the stroller. He willingly hands me $150. Faith is able to extricate her gear from the wreckage, loads it onto one of our vehicles, and continues marching.

Since the start of the March my biggest fear has been that one of us will be struck by a car or truck. I think of how many vehicles have passed us since Los Angeles. Hundreds of thousands? Over a million? It's impossible to say. I think, too, of how America's highways are so unwelcoming, if not downright hostile, to those with the audacity to travel on foot. It's almost incredible that Faith is, so far, the only marcher to be hit. Perhaps her close call will wake others up to the importance of caution, especially as the roads ahead will be some of the busiest and most dangerous yet.

At our campsite in Elmore, several local residents visit, including a man who lives nearby. He invites marchers to use his shower. Lala and I are the only two marchers interested in taking him up on the offer. We walk to his apartment separately. I arrive first and find a half-finished construction project on the second floor of an old brick building. The repair work seems unnecessarily chaotic and undirected. There's something odd about it, and the guy makes me nervous. I finish my shower and wait for Lala. I have visions of the man being some demented creep who might do something horrible to a woman alone in his apartment.

When Lala finishes her shower, she's surprised that I'm waiting but noticeably relieved. As we walk down the stairs toward the street she whispers to me that the guy made her nervous, too. She thanks me for staying behind. For the first time in weeks, we talk without arguing, exchanging casual conversation on the way back to camp. Lala tells me that she and Jimmy are no longer "seeing" each other. I nod in acknowledgement as I consider what affect this might have on the March's power struggle. Perhaps favorable? I'm not sure.

At our morning circle on September 29, Sarah announces, "We have an easy day today, just 14.3 miles. But the next five days will average 20 miles each." There are groans as Sarah continues, "Sorry, but you all wanted to go to New York, so this is what we've gotta do to make it to DC by November 1."

Unexpectedly, the string of long days takes a harder toll on me than any other marcher. I start a little late on October 3, a 17-mile day from Oberlin to North Olmsted, a suburb west of Cleveland.. I expect I'll have no trouble catching up with others. But since Toledo, plantar fasciitis has returned. My feet hurt more and more as the miles pile on. I eventually catch up with some of the slower marchers. Miriam treats me to lunch and we walk together for a ways. The cumulative punishment of seven months of marching weighs heavier and heavier on her. "Don't wait for me," Miriam says. "You don't need to go slow on my account."

"Today, I'm the one trying to keep up with you, I'm afraid," I tell her. "That plantar fasciitis is back and it's been getting worse all day. In fact, I've gotta take a break. You go on ahead and don't let me slow *you* down."

"Well, how about that for a change," says Miriam. "Sorry to hear it. I'll let Sarah know. If you aren't able to make the last couple miles, I'm sure someone will come get you."

It's been raining for the past hour. My torso is dry but my legs are soaked. We pass a roadside shelter where I say goodbye to Miriam and stop to rest my feet. I sit at an old wooden picnic table soiled with the stains of spilled soda and greasy food. I position myself with the wind at my back and try to stay warm while massaging my feet. The massage helps, but after ten minutes, my wet legs grow cold and I need to get moving again.

The last two miles are excruciating. My feet are throbbing painfully as I arrive at our host church in North Olmsted. After supper, I find a quiet room with a few other marchers and crawl into my sleeping bag early. Our wake-up time is 6:00, and I hope twelve hours of rest will make a difference. I drift off to sleep uneasily, recalling what Charles said when plantar fasciitis first surfaced in Nebraska: "If it keeps getting worse, which is likely, you may be forced to quit whether you want to or not."

CHAPTER 37

RESILIENCE

"The greatest lesson I learnt from him was to be fearless. He would always tell me, 'If you are fearless, there will be no difficulties in life.'"

— SUMITRA KULKARNI,
WRITING ABOUT HER GRANDFATHER, MAHATMA GANDHI[1]

Simple pleasures matter more when life is hard. Lately, I find uncustomary comfort in the meager arsenal of possessions known as "mine." The feel of *my* walking stick cupped in my hand, the strap of *my* satchel pressed against my shoulder, the coziness of *my* tent at the end of the day, *my* phone's "Fairy Fountain" alarm — a gentle, watery melody that each morning brings a smile to my lips as it vanquishes the last lingering strands of whatever dream has amused or plagued my subconscious.

Today when "Fairy Fountain" goes off, I sense something is wrong. The phone is charging at an outlet fifteen feet away. I'm groggy and confused, then recall setting the alarm for 4:00 a.m. two days ago when it was my turn to cook breakfast. "Shit! I forgot to turn it off last night!" I mutter to myself.

The alarm wakes everyone in the room and, for once, the fairies and their fountain don't make me smile.

I hastily roll out of my bag and try to stand, but my legs crumple underneath me. I whisper a couple choice words as I drag myself to the

---

[1] http://www.penmai.com/forums/personalities/14887-sumitra-gandhi-kulkarni-grand-daughter-mahatma-gandhi.html

phone, turn off the alarm, then crawl back to my sleeping bag. For the next two hours I lie awake, angry at my body's sudden and dramatic failure and wondering what, if anything, I can do to keep walking.

Fernando helps me pack my gear and loads it onto the truck. Kathe finds a spare set of crutches that Mack tries to adjust for me. The armpit pads are badly corroded. One of the stick tips is missing, so each crutch is a different length. I test the crutches in the parking lot but can't imagine walking even a few blocks like this, let alone 21.4 miles to our destination in Cleveland Heights.

I give up on the crutches and decide to hobble along as best I can. It's a long shot, but if I walk half the distance today perhaps I can walk the other half tomorrow on our day off. I pop a few pills to dull the pain and set out thirty minutes before other marchers.

I move more slowly than ever before. I shorten my stride and walk on grass or dirt whenever possible. I rely heavily on my walking stick, planting it firmly in front of me and pushing so hard that it bends a bit. The stick and my arms absorb much of the impact. With these adjustments and ibuprofen, the pain in my feet is at least manageable.

Other marchers catch up as I complete two miles. A few offer to walk with me. I tell them I appreciate their kindness but don't want to inconvenience them. I can always call Sarah if I can't continue.

After the last marcher passes I stop at a park bench. The day is cool but comfortable, with an occasional light mist of no consequence. I remove my shoes and socks and gently massage my feet. I'm surprised at how good it feels and continue rubbing my feet for ten minutes. When I resume walking, I find that the massage has further reduced the pain.

I stop every hour to repeat the treatment, regardless of how socially awkward the circumstance. I massage my feet while sitting in bleachers watching kids' play soccer. I stop at an Irish pub where I have lunch while rubbing my feet under the table. The pub doubles as the headquarters for political candidates with Irish last names. It's a busy, bustling place, and if staff or patrons notice me in the corner, massaging my bare feet between bites of shepherd's pie and boxty,[2] they say nothing.

With the punctual discipline of a cloistered monk chanting regularly throughout the day, I religiously administer my hourly foot massage — once sitting on a wall, once on a guardrail, once on a large

[2] https://en.wikipedia.org/wiki/Boxty

rock. It takes me seven hours to finish the first eleven miles — a distance I would normally knock out in half that time. At this pace, I won't get to Cleveland Heights until 9:00 or 10:00 tonight. But I feel I can keep going, complete the day, and benefit from tomorrow's rest day.

I plod along, stopping every hour, far behind other marchers and the two support vehicles that no longer scour the route for laggards. Sarah calls to check on me and offers to pick me up. I tell her I think I can make it. She calls me stubborn and we both laugh.

I walk in the dark for the last five miles. Part of the route is through the Scovill Avenue Neighborhood, which I later learn is one of the most dangerous places in America. An ABC News report rated Scovill as second in terms of neighborhoods where people are "most likely to be murdered, raped, robbed, or have their car stolen."[3] The report lists my chances of being a victim as one in six. It's just as well that I walk in ignorance, unaware that local supporters had expected us to stay together through Scovill during daylight hours.

My massage break in Scovill is on an upended five-gallon bucket across from an abandoned building. The facing wall is adorned with graffiti, painted by someone with evident talent and abundant access to paint. I admire the colorful, flowing patterns and letters, illuminated in the yellowish glow of a weathered street lamp. I try to imagine who the artist might be, what they intended to say, what they do for a "day job," and if they've ever been arrested for practicing their art.

I get wrapped up in the possibilities, and start talking out loud to myself.

Ed 1: "So, Ed, why is this artist, if caught, likely to be sent to jail while Malevich's 'Supremo Composition,' which to my eye is a confusing assortment of meaningless rectangles, sells for $66 million?"[4]

Ed 2: "Good question, Ed. And what distinguishes prized art from vandalism? Is it the income level of the artist, the prospective buyer, or maybe both?"

Ed 1: "Well, Ed, it's income for sure, but it's also the art's canvas. Graffiti confounds and infuriates the pathologically wealthy because they can't capture it, own it, make it 'mine.'"

---

[3] http://abcnews.go.com/Travel/LifeStages/americas-dangerous-neighborhoods-areas-violent-crime/story?id=11803334

[4] https://en.wikipedia.org/wiki/List_of_most_expensive_paintings

Ed 2: "Great point, Ed. Art on the side of a building, a train, a retaining wall — you can't haul that home to your mansion. Even if your palace is big enough, those canvases would clash with the wine cabinet, the cactus table, and the tiger-skin rug."

A handful of people pass while I'm talking to myself and massaging my feet. A few glance in my direction. Most don't seem to notice me let alone have any interest in my perspective as an art critic.

I set out on the last two miles. Most of the people I pass look right through me as if I'm wearing an invisible cloak. Whether through magic or dumb luck, I avoid joining the ranks of the 16.6 percent of Scovill visitors victimized by violent crime.

This morning, when I started the day dragging myself across the floor because my feet hurt so badly, I never imagined I'd be able to walk 21.4 miles. As I arrive at my destination in Cleveland Heights, I feel an incredible sense of accomplishment. Innovation and determination has carried me through a fourteen-hour day — the most difficult day yet.

Inside the church, Sarah greets me with a plate of food she saved from supper. Yes, simple pleasures matter more when life is hard.

Our day off on October 5 does my feet more good than I could have imagined. The next day's 17-mile march to Macedonia is uncomfortable, but I stop less often and I'm able to keep up with other marchers.

October 7 starts with unexpected contention. There's a general understanding that our morning circles will be brief and focus strictly on the day's route and scheduled activities. Today, a few marchers insist on sharing ideas they've come up with for direct action when we arrive in Washington, DC. They tell us the discussion will take about twenty minutes. I object, but others want to hear what they have to say.

John Abbe leads the discussion. The plan is to target the Federal Energy Regulatory Commission (FERC). "If even one marcher objects to a suggested action we'll remove it from the list," says John.

I glance at my phone and set the timer for twenty minutes. "Ok," says John, "the first idea is to superglue ourselves to security officers."

"Wait, is this a joke?" I ask. The expression on John's face confirms he's serious. "Do you not realize what a horrible idea that is?" I say.

"Do I take that as an objection?" John asks.

"Yes, emphatically," I reply.

"We'll remove it from the list," he says. "Ok, next idea. We know someone who has access to Renaissance Faire equipment and they can get us a catapult. We can use the catapult to launch marchers over FERC's wall and into the compound."

"What!?" I practically scream. "Are you out of your mind?"

John just looks at me as I say, "Object."

"Ok, gunking up the locks of doors on the FERC building," says John.

"Object."

"Dumping toxic waste or human feces in the lobby."

"Object!"

"If it's cold enough outside, spilling water in front of the building so it freezes and people have a hard time getting in."

"By 'have a hard time getting in,' I assume you mean, 'fall and injure themselves'?" I ask.

Again, John just stares at me. "Object," I reply.

Every so often, John suggests an idea that's not over-the-top batshit-crazy and I don't object. There ensues much conversation that drags on, eating up big chunks of time.

After 20 minutes my alarm goes off but John's not finished. "I'm leaving," I announce. "Who's coming with me?"

Miriam and Fernando step out of the circle. Miriam is frustrated, too, and says, "I feel like the March has been hijacked."

The three of us head out. I'm embarrassed that we have guests this morning and surprised that none of them nor other marchers leave with us. I think of the proverbial train wreck: you don't want to watch but it's impossible to look away.

Tonight in Kent, I stay at the home of Silvia Rhodes and Landon Hancock. Landon is the associate professor at the Center for Applied Conflict Management at Kent State University. The Center was established in 1971 in response to the killing of four students by Ohio National Guard soldiers during a protest against the Vietnam War.[5]

I share with Landon the crazy ideas marchers suggested earlier today. "Yeah, gluing oneself to a police officer isn't exactly peaceful or nonviolent," Landon laughs. "Sounds like some of your marchers could benefit from our program."

Landon describes how the university has become widely known for its focus on peace and conflict resolution, though many people still associate Kent State with the tragic incident in 1970.

"It's a lot easier to understand nonviolence in theory than to implement it in practice," I offer. "I've studied Gandhi extensively. His granddaughter is a friend of mine, gave me this satchel, in fact. I even went to India in 1995 to meet people using Gandhi's ideas to address various injustices."

Landon's intrigued, and I tell him more about my trip to India and what I learned. "But no matter how well I think I understand nonviolence, hardly a day goes by on this March when I don't screw up," I confess. "I can't tell you how many times I've gotten angry or sworn at someone. It's embarrassing and hardly nonviolent. You'd think I'd have a better handle on this by now."

"It's certainly easier to preach nonviolence than to practice it, especially when you're doing something as physically and emotionally intense as your March," says Landon.

We talk for an hour or so about the importance of continued study and dialogue. I hope Landon's insights will help me do a better job at managing my own temper during the final few weeks of the March.

---

[5] https://www.kent.edu/spcs/profile/landon-hancock

A MEMOIR FROM THE GREAT MARCH FOR CLIMATE ACTION

CHAPTER 38

TENT

"Everyone knows that fracking poisons the air and water. We wanted to show how it tears apart local communities and subverts democracies and corrupts political leaders and eviscerates all the things that Americans value."

— MATT DAMON ABOUT HIS FILM ***PROMISED LAND***

Our final day in Ohio is a 21-mile march to Youngstown, recently lauded as "America's fastest shrinking city."[1] The town's population peaked at 170,000 in the 1930s. Today, it sits at 65,000.[2]

I start early and walk slowly, concerned that a fast pace on a long day might rekindle serious foot problems. Since Cleveland, my feet have been tender but reasonably cooperative. I hope I can maintain this status quo for another three-and-a-half weeks.

It's refreshing to be on the road again before sunrise. Occasionally, I glance back at the moon sinking in the west as I walk toward a celestial canvas of red and orange veins that grow more brilliant over the next hour. I walk on grass as much as possible, but the dew persists well into the day and my shoes and socks are drenched by mid-morning.

Between my slow pace and frequent stops, other marchers catch up a few miles outside Youngstown. I walk in silence with Sean, Mack, and a new marcher, Annamarie Chantel, through a tortured landscape of industrial decay, a testament to the failure of humanity's embrace of unsustainable technologies.

We average only 14 miles each of the next five days, though

[1] http://www.hamptoninstitution.org/youngstown.html#.Wj27OLaZOb8

[2] http://population.us/oh/youngstown/

frequent steep elevation changes minimize the advantage of shorter distances. The rolling foothills of the Allegheny Mountains are a welcome change of scenery, but amidst the beautiful hills and forests are fields stripped bare by fracking. Towering drill rigs replace century-old oak trees. Gas flares shoot red and yellow flames where thick stands of maples once colored the autumn sky. Many people we meet share stories of how fracking has compromised or even destroyed their lives. Those conversations often end in tears and hugs.

On October 11, local supporters organize an anti-fracking rally in Butler, Pennsylvania, population 13,000. They hire a bus to pick us up near our campsite in Darlington, an hour away. Our driver pulls in to Butler just as the rally begins. Stepping off the bus, we're greeted with a heroes' welcome as 150 supporters cheer and three TV crews swarm us. One reporter says to me, "You guys are kind of like rock stars here."

It's encouraging to know our efforts are appreciated, and important to be reminded that our steps and sacrifice matter. At the same time, I feel sad that little Butler hosts the only rally across Indiana, Ohio, Pennsylvania, and Maryland combined.

Afterward, the Mountain Watershed Association treats us to spaghetti and meatballs at the American Legion hall. I thank the Association's director, Stephanie Novak. Over supper, she and I discuss the March's tight budget, and she offers to expedite a $2,000 donation to cover food expenses over the next two weeks.

Arriving back in camp, the night quickly becomes cold. I appreciate the occasional home stays or floor space at a church. Tonight, I'm happy to crawl into my tent. It's become my refuge, my home. It's bedroom, sanctuary, and womb all in one.

Like me, my tent bears the scars of a long, long march. The zipper broke months ago. But with safety pins and duct tape, I'd jerry-rigged that zipper to keep rain, wind, and mosquitoes at bay — mostly.

The walls of my tent are as thin as plastic bags. Yet they provide not only protection from the elements but comfort, security, and privacy — or at least the illusion of privacy.

My tent is my friend and safe place. It's one of the few features that remain rock solid in the fluid life of a cross-country marcher, where turmoil and discomfort are the norm.

My tent and I weren't always friends. Sure, there was the honeymoon. For the first two weeks, sleeping in her was a novelty, fun even — except during four days of torrential rain that transformed my doormat into a mud puddle.

But the honeymoon passed and my tent became a burden, an annoyance. In the desert, we fought all the time. Even with Shira's help, setting her up and taking her down was a daily drain on my patience and energy.

But as the physical and emotional trauma of the March grew worse, my tent and I drew closer. She became my best friend, the one place I could always count on for peace and comfort.

Once inside my tent, I didn't want to leave until morning. Initially, that desire conflicted with the need to pee. But somewhere in the desert, I acquired the knack for suppressing the pee reflex, assuring an uninterrupted sleep.

Well, ha! If only that were true. A full night's sleep has been rare, and external interruptions frequent. Trains are the most common, with traffic a close second. But the diversity and unpredictability of things that make noise in the night — a gang of roosters, a truck spraying for mosquitoes, ducks having sex, high winds that roared through the walls of the tent with a ferocity that made sleep nearly impossible.

But the greater disruptions to sleep often come from within. Perhaps it's my near-constant fatigue, but dreams in my tent are more vivid, memorable, and frightening. Lately, they often involve women.

On this particular night in mid-October, when the dew has settled thick on the grass of our campsite in this small Appalachian town, the image of a woman I had loved prior to Grace enters my subconscious in a harrowing blend of reality and fiction.

She appears before me. We only know each other casually, so we approach with caution. We embrace. I feel my knees go weak, my arms tingle. We cling to each other tightly as our common bond of loneliness kindles a fire smoldering in our hearts. Our breathing becomes labored. I feel myself slipping into unconsciousness.

When I wake in my dream, we're lying in a bed draped with an elaborate canopy of lace and lights. Fruits and vegetables hang from the canopy. Some are fresh and fragrant. Others are rotting and starting to smell. The canopy drips with honey. A trace of nutmeg wafts through the air.

My lover and I are consumed by lust. The juices of sex splatter around the bed and onto the canopy, blending with the honey, covering the fruits and vegetables. It's wonderful, surreal, tantalizing, yet deeply disturbing.

The sex turns violent. My lover screams at me, berates me. She threatens to poison, stab, and hurt me. I realize she's not my lover but some monster, some creature spewing love and hate in the same venomous breath. If I linger, she'll devour and destroy me.

I rush out from under the canopy and flee down a long, dark tunnel that twists and turns. Every time I glance behind, she's there, staring at me, her eyes growing fiery and hollow, burning like two hot coals with both passion and contempt.

I run faster, deeper into this labyrinth, turning one way then the other. Though I feel I've traveled a great distance, she's still there, always there, sometimes smiling kindly, sometimes glowering like a mythological creature sent by the gods. Why? To torture? To taunt? To teach? For what purpose, I don't know. I don't want to know.

I wake up. I have to pee. This time, I'm grateful that the pee-suppressing reflex has failed. I scramble on all fours out of the tent that at this moment feels more like tomb than womb.

The dew is now a thick layer of frost on the grass, glimmering like so many crystals in the faintest streaks of moonlight. My feet are instantly and thoroughly soaked as I walk through the tall, luminescent grass, looking for just the right place to pee under a cold mountain sky.

I don't mind wet feet. I'm happy to have escaped my tent, happy to pee, eager to purge my body of whatever toxins are responsible for such dreams, eager to forget.

The march from Aliquippa to Ben Avon presents a challenge we haven't experienced since Payson, Arizona. Our Pittsburgh hosts are adamant that a 3.3-mile stretch of Ohio River Road is too dangerous to walk. They tell us there's no alternative route and marchers will have to shuttle.

Mel Packer, a former truck driver, says, "Ohio River Road is out of the question. It's a four-lane highway with no sidewalks and so narrow that us truckers would sometimes knock our mirrors off on the telephone poles. *No one* ever walks or bikes on Ohio River Road."

I tell Mel and others that I appreciate their concern, but skipping part of the route is not an option, especially for those who've walked every step of the way.

My primary contact in Pennsylvania is Stephen Cleghorn, who's worked hard to help us with campsites, potlucks, routing, and promotion. Among other contributions, Stephen helped convince Jackson Browne to give us a couple shout-outs at a concert in Pittsburgh last night, with Jackson saying as he left the stage, "Don't forget to join that climate march!"

Stephen's been a great ally, but like Mel and others, he doesn't want us walking on Ohio River Road. I explain that we've become masters at negotiating such conditions and that it's symbolically important for Miriam, Jeffrey, Mack, and me to maintain our trail of unbroken steps.

Stephen agrees to broker a compromise. He talks with other local volunteers who grudgingly accept that four of us will walk Ohio River Road if all other marchers board shuttles.

Thirty marchers arrive early afternoon at the junction of Beaver Street and Ohio River Road. A couple dozen local supporters are waiting. Some look concerned. There's palpable tension and an awareness that something risky and unprecedented is about to happen. Some again try to dissuade us. One even warns that walking on Ohio River Road is almost suicidal. I assure them we'll be extra careful and that we're adept at managing heavy traffic. Trying to lighten the air I add, "Besides, drivers know if they hit us their insurance rates will skyrocket, and no one wants the hassle of scrubbing a splattered climate marcher off their hood."

But there's a knot in my gut. I think of Faith being hit just two weeks ago. I think of the blown tire that grazed Steve's head. I think of the many times drivers have buzzed us and wonder if I'm making the right call, or if pride is blinding me to the risks.

Though I'm no longer in charge of this "leaderless" collective, everyone defers to me on the decision. I approach Miriam, Jeffrey, and Mack privately and ask them how they feel. All three are emphatically ready to give it a shot and don't want one short stretch of highway to break our string of 2,900 miles.

I adjust my safety vest and we set out. I lead the way with Miriam behind me. Mack is next on crutches and Jeffrey brings up the rear. With two lanes of traffic and a steady stream of vehicles, my papal wave is of little use. At first, there's a sidewalk — narrow and in bad repair, with chunks missing and weeds growing over it in places. The sidewalk soon disappears and we're sandwiched between vehicles and vegetation. Occasionally there's a cliff to our left, and it feels like we're running a gauntlet between rock and metal. At one point we negotiate an off-ramp and an on-ramp.

It's clear why pedestrians and bicyclists avoid Ohio River Road. But as seasoned veterans of highway walking, this doesn't seem like that big a deal. Halfway through, I cock my head and yell back to my companions, "Well, this isn't so bad, is it?"

"No, not at all," replies Jeffrey. "In fact, it's kind of anticlimactic after all the hype."

"Yeah, what a let-down," Mack jokes. "I was looking forward to a real challenge, something truly death-defying."

"Oh, listen to you macho guys," Miriam chimes in. "You've got a point though. It's not a road I'd want to walk every day, but they sure made it sound worse than it is."

We jest a bit about wishing the road were more challenging, then walk again in silence. But there's truth in our jesting. Seven months ago, Ohio River Road would have terrified me. Now, it's strangely exhilarating. Acknowledging that I'm even a tiny bit disappointed that the road didn't live up to the level of danger we expected feels wrong, demented even. I want to ask Miriam, Jeffrey, and, in particular, Mack, if they feel something similar, but I'm embarrassed to even pose such a question.

A couple days ago, Fernando and I were negotiating a particularly bad stretch of highway when he said, "If I were the kind of guy who found death arousing, I'd have a huge boner right now." I laughed at the time, but perhaps Fernando was on to something. We fear death when it's mysterious and unknown. When we face it repeatedly in an arena that's become familiar — in our case, as gladiators battling four-wheeled assailants in a four-lane coliseum — in place of fear we discover exhilaration, not so much at the prospect of death, but at the beautiful, tenuous gift of life.

CHAPTER 39

PUZZLE

"I loved you in the morning, Our kisses deep and warm. Your hair upon the pillow, Like a sleepy golden storm... But now it's come to distances, And both of us must try. Your eyes are soft with sorrow, Hey, that's no way to say goodbye."

— LEONARD COHEN, "HEY, THAT'S NO WAY TO SAY GOODBYE"

The next day, Mack, Miriam, Jeffrey, and I walk together through downtown Pittsburgh. It's lunchtime on a Tuesday and foot traffic is thick. We're an odd foursome: Miriam with her yellow safety vest and purple water bottle, Jeffrey sporting a thick beard and cap that suggest the fusion of Fidel Castro and Leo Tolstoy, Mack on crutches with his wildly untamable hair, and me toting my walking stick and satchel. Most days, climate marchers are an object of curiosity. Today, we hardly land a glance. I joke that it's refreshing to be where strange-looking people are so abundant we blend in.

We walk through Schenley Park and come to a bridge spanning a deep ravine. Across the bridge runs a fence covered with hundreds of locks. Many are inscribed with names and initials: Blake & Amara, KW & BZ, JM loves KH. A woman walks by and I ask her about it. "These are love locks," she tells me. "Couples put them here to show their commitment — you know, like they're 'locked' together."

I'm fascinated and later learn that this sweet practice began over a hundred years ago in a small town in Serbia. A young school teacher named Nada allegedly died of a broken heart after the man she was engaged to abandoned her. Other young women, hoping to avoid a similar fate, wrote their names and their man's name on padlocks and fastened them to the bridge where Nada had fallen in love.[1] The practice spread around the world and is now popular on several of Pittsburgh's 446 bridges.

We take a few photos then continue to our end-point at Church of the Redeemer. I want to snuggle into my tent, but that's not an option tonight. It's either communal lodging at the church or home stays — for me, an easy choice.

My host family is pleasant and hospitable. Their German Shepherd, on the other hand, wants me dead. This demon of a dog growls fiercely every time he sees me, baring fangs that snarl "dismemberment" in Canine. When the monster isn't out in the yard, I'm a prisoner in my basement bedroom.

But I don't mind. The bedroom is quiet and comfortable, with two small windows providing a little light and air. The bed is huge and covered with a fresh, white comforter. I spread my arms and legs across its vast expanse and imagine myself lying on a frothy cumulus cloud. I can't get the love locks of Schenley Bridge off my mind. I wonder if the commitments sealed by this ritual boast a higher success rate than the ritual of marriage, which in America fails about half the time — or in my case, 100 percent of the time.

Of life's many conundrums, none confounds me as thoroughly as love. With Grace, I was convinced she was the woman I was meant to spend the rest of my life with. When Grace and I were together — conversing, walking, eating, making love, or just sitting still — I felt a profound sense of wholeness. She seemed to be the piece that completed the puzzle of my existence, the answer to the riddle of human loneliness in this unfathomably vast, hungry, and compassionless cosmos.

Yes, everything felt right when I was with Grace. The trouble was we were rarely together. For every ten activities I'd propose, one would pan out. She'd frequently cancel planned events, often at the last minute. Once, she was inexplicably two hours late. If I

---

[1] https://en.wikipedia.org/wiki/Most_Ljubavi

hadn't been so blinded by love, perhaps I would have noticed that the puzzle piece I thought fit so well was being forced into an empty space where it really didn't belong.

When the March came through Des Moines and Grace ignored me, I finally noticed, finally understood. After I wrote my farewell letter, she called almost every day. She seemed genuinely remorseful, and I began to question whether I had acted hastily.

Then Grace said she'd visit in Chicago, and didn't. She said she'd visit on another occasion, and didn't. In Indiana and Ohio, her phone calls became less frequent. I began to feel relieved not to hear from her — or so I told myself.

I want to believe that I'm over Grace, but I wonder what she's doing tonight. Does she miss me? Perhaps she's lying on a similarly expansive and lonely mattress. Perhaps she's moved on and is lying in bed with another guy. I want to call her, but don't. I remind myself that we have a full day tomorrow and I need my rest.

I burrow into my cloud and sleep comes quickly. Dreams of women again plague me, one after another in rapid succession. I'm in a room crowded with people who prattle on about mundane matters. I'm bored by the banality of the conversation when I hear a woman's voice rise above the cluttered chatter. "What's this?" I ask. "An intelligent thought? An observation on a matter of substance?" My eyes and ears are drawn to her. I recognize her instantly: My soul mate! I walk toward her across the room, but she laughs at me and turns toward another man. They kiss. Then suddenly, tragically, her body disintegrates before my eyes as I recoil in horror.

I run fast and far, halfway across a continent, until I'm walking along a river next to a woman. It's raining and cold. The wind howls in the trees above us. To escape the fury of the storm, we retreat to a room. We fall into bed and make love for hours and hours — never eating, never sleeping. It's the most exhilarating sex either of us has ever enjoyed. We're satiated, ecstatic, exhausted. Then we hear a small, thin howl in the distance, a wailing voice filled with sorrow. It fades, then vanishes completely. I look at my lover and she, too, begins to fade — slowly, painfully, with a sad, resigned smile until she's gone.

I'm now in a cabin in the forest. Six people are sitting around a table, sharing food and conversation. A woman says, "All I really want is to live in this cabin, raise animals, and grow my own food."

"Ah!" I say. "My soul mate!" She smiles at me. She invites me to her bed and falls asleep immediately. All night long, I lie there with an erection so large it hurts, as if the skin on my penis might tear if it grows any bigger. I panic as I realize that if it does, my penis will be chopped off by the blades of the ceiling fan above the bed. I try to relax, try not to feel aroused, all the while concerned about the fan that seems to rotate faster and faster — and imperceptibly closer and closer — as the night wears on. Eventually, the first light of dawn appears and the woman awakes. She looks at my erect penis, laughs, and apologizes halfheartedly for letting me believe she was my soul mate. She then jumps out of bed fully clothed and walks away with two of my male friends, each holding one of her arms. I feel humiliated, but say to myself, "At least you still have a penis."

I wake with the entire dream sequence clear in my mind, too clear, too vivid, too disturbing. "What's the message here?" I ask myself groggily as I crawl out of bed. I check my phone. It's only 11:00, yet I feel like I've slept for hours.

"What's the lesson, Subconscious? Come on. You got something to tell me, or are you just messing with my head?" I say, genuinely pissed off that I'm having these dreams.

I pee. I want to walk around the basement for a bit but remember the canine jaws of death that lie in wait beyond my door. I walk in circles in the room for a few minutes. I miss my tent. At least there the disturbing dreams lunge at me one at a time. The austere comforts of a tent, coupled with the exhaustion of twenty-mile marches, leave little room for indulgence in the frivolous affairs of the heart. On a big bed in a comfortable room, I can't help but think of Grace. The bed drags me back to the times we made love. I want her so badly. Yet just as badly, I want nothing to do with her ever again. Sure, I was the one who ended our relationship two months ago, but only after she ended it a hundred times through so many acts of indifference and rejection.

Karma, I tell myself. That's what this is. There was a woman in my life years ago, a woman who was certain that I was her soul mate. I now know how she must have felt, perhaps still feels. I pull the big white comforter off the bed, fold it in thirds, and lay it on the carpet. I drag a pillow and a blanket onto the floor with me and curl up on this makeshift mattress where I sleep dreamless for the rest of the night.

A MEMOIR FROM THE GREAT MARCH FOR CLIMATE ACTION

CHAPTER 40
FORGIVENESS AND REDEMPTION

"Redemption comes to those who wait, forgiveness is the key."

— TOM PETTY, "LONESOME SUNDOWN"

Trudging through the Appalachian Mountains for twenty miles a day is hard on the legs and even harder on the knees. The steep descents prove especially difficult for Miriam. I lend her my walking stick to help manage one particularly long stretch.

Despite the challenging terrain, marchers are excited about our pending arrival in Washington, DC. I talk frequently with DC Action, our organizing team there. They've now secured permits to march in the street and for our rally across from the White House in Lafayette Square.

Unbeknownst to me, a group of marchers led by Lala has been quietly building support to hold a vote on who gets to speak at the rally. By the time I learn of it, they've already scheduled a meeting to conduct the election. I approach Lala and ask, "Have you considered why this might be a really bad idea? There are solid reasons why organizers don't hold an election to choose rally speakers."

"Everyone deserves a voice regarding who gets to speak," she explains. "Our plan is to take nominations and then elect three speakers."

"That makes as much sense as voting on campsites or what's for supper," I say. "And it undercuts the work our people in DC have been doing these past three months."

"Marchers want more input," insists Lala. "They're tired of meaningless rallies. And to be clear, *only* the three people we elect will get to speak."

I hear what she's saying. This is yet another power grab, a clever attempt to prevent me from speaking at the rally. I stare at Lala for a few seconds, shake my head, and decide to save further commentary for the meeting.

I hurry through the next day's march, walking faster than I should in order to make the 2:30 meeting. When I arrive, I'm told the meeting won't start until 3:30. At 3:30, it's changed to 4:15. It eventually starts at 4:45. I say nothing but I'm seething, knowing I could have walked slowly enough to avoid the risk of inflicting further damage on my feet.

Lala facilitates, of course. At times the gathering feels more like a circus than a serious meeting. Berenice Tompkins nominates one marcher after another, saying whoever's name pops into her head. About half the marchers, including me, are nominated. Each then gives a speech about why they want to give a speech.

When it's my turn, I stand and say, "This is a really, really bad idea. When you're sick, you see a doctor. If your toilet's clogged, you call a plumber. You want piano lessons? You hire a pianist, not a cellist. Like other professions, organizing is a skill set. If you want a great rally, you don't hire a doctor, or a plumber, or a musician. You hire an organizer, someone with ability, training, and experience.

"A few of you think our rallies have been meaningless. I find that insulting — not only to our staff but to our local allies who've worked tirelessly to help pull together not just a rally but the route, campsites, potlucks, media interviews, and more. Our local allies are the ones who understand their communities. They know how to frame a message that works. They know how to line up speakers so key constituencies are represented. In my thirty years of organizing, I can't think of a single successful rally where people elected the speakers."

I pause and consider whether I should say what else is on my mind. I shouldn't, but I say it anyhow. "At most of our rallies, I come to the mic simply to make an appeal for funds. In Los Angeles and Des Moines, I spoke from my heart about why I felt compelled to organize this March. I've always planned to speak in Washington, DC, too. So on November 1, in front of the White House, after I've walked over

3,000 miles for climate action, I'll give my speech whether you vote for me or not."

The meeting drags on for over three hours. In the end, marchers vote for three people and I'm one of them. The other two speakers are also white and male.

The next morning, I point this out to Lala. "Here's another reason you don't vote on who gets to speak at a rally. Marchers just elected a slate of speakers with no gender or racial balance. How do you feel about that?" I ask, feeling more than a little smug.

"Oh, we fixed it," says Lala.

"What do you mean you 'fixed' it?" I ask, genuinely perplexed.

"People agreed that we couldn't have only white men speak at the rally, so we changed the line-up of speakers."

"Um, I'm all for diversity, but you held an *election*. People *voted*. You yourself collected and counted the ballots," I remind her. "Do you know what it's called when you toss out an election because you don't like the results?"

"What?" asks Lala sharply, increasingly offended by what is, admittedly, a condescending lecture.

"It's called 'fixing' an election," I say as Lala begins to walk away. "Fixing an election is the antithesis of democracy. It's the hallmark of a dictatorship," I say, my voice rising as Lala moves out of earshot. "The United Nations sends monitors to unstable countries to guard against people like you," I yell after her.

I want to add that if I'd known she was going to do this, I would have secured independent observers to protect us against this subversion of democracy. But Lala's gone, and that's just as well. She got her vote. I'll still speak at the rally. Everybody's happy, damn it, happy.

As I get ready to walk, my gut says this won't be the last volley between us in a power struggle that must look downright comical to independent observers — UN-trained or otherwise.

These days, I often walk with Fernando. Our conversations bounce around from serious to silly with little regard for political correctness. Walking into a small town today, we're surprised at the complete lack of any human activity. "Folks are probably peeping at us from behind their shades," says Fernando.

"What do you suppose they think of a scrawny Irishman walking into town with a big ugly Indian?" I ask.

"Oh, they probably think you're bringing me in for bounty," says Fernando. We laugh for most of the next block. "Now they see us laughing like fools and think we just escaped from the loony bin."

We stop for lunch and gulp down 16-ounce ribeyes, home fries, and an entire loaf of fresh-baked bread slathered with butter. I'm amazed at how much we need to eat to keep pace with the caloric demands of marching. "Given what we pack away, I bet our carbon footprint is worse than if we drove," I suggest.

"Yeah, let's ditch this marching shit, steal a car, and drive the rest of the way," says Fernando. "Do it for the planet."

After lunch, our conversation turns more serious. As with Steve, the March's dysfunction wears heavily on Fernando. Unlike Steve, Fernando rarely holds back. Yesterday, Lala assigned him his job for the week. He told her to fuck off, said he had no respect for her, and that he'd do whatever job he felt like doing.

"Yeah, Lala gets under my skin, too," I say. "She's got a knack for that. I don't handle her well either, but you might have me beat."

We talk about some of the struggles we've faced in the past. I tell him about the time my parents signed me up for swim lessons. On Saturday mornings, they'd drop me at the Lynn Boys Club where a large, gruff man taught a class of about 25 pre-adolescent boys. In the locker room, he told us to get naked, insisting it was easier to swim without clothes. Nearly all the kids complied, except two other boys and me.

The cold, chlorinated pool of naked boys was unnerving — the instructor, terrifying. He'd walk around the pool yelling at us, poking boys with a long stick. I'd cower near the edge of the pool, never daring to swim, hoping to escape detection for one harrowing hour each week.

One day, he came over to me and said, "You *have* to learn how to swim. You're not even trying. Start by just putting your head in the water and blowing bubbles." I looked at him suspiciously. He must have read my mind because he said, "I promise I won't hold your head under water."

I took a deep breath, cautiously lowered my head into the pool, and as soon as I started blowing bubbles he grabbed my neck and held me under water. When he let go, I came up gasping and choking, terror in my eyes, the burn of chlorine in my throat.

"Wait, he did *that*, and you didn't tell your parents?" Fernando asked incredulously.

"I guess I was taught never to question adult authority," I said. "So yeah, even after that incident, I said nothing to my parents."

"No wonder you're a fucking mess," Fernando says.

"I'm not a fucking mess, you're a fucking mess," I shoot back with a grin.

Fernando laughs. "Yeah, we're all damaged goods. Seriously, though, that's some bad shit. Prison's too good for that creep."

"Yeah," I respond. "If there's a God, the guy's in Hell, naked and flailing about in a cold tank of foul water, some snickering devil poking him with a stick every time he comes up for air."

We stop to admire a scenic overlook, taking in the incredible panorama of the Allegheny Mountains. Looking closely, I make out scars inflicted by coal mining and fracking. But where the mountains are free of humanity's meddling, they speak confidently of the dignity and perfection of nature.

"It's hard to talk about human ugliness when you're looking at that," Fernando says. "But I'll tell ya, I had plenty of rough spots growing up, too. My Mexican father abused alcohol and drugs. He'd park in an alley and leave me sitting in the car while he went off to shoot heroine for twenty minutes. My uncles were wild, too — bikers, fighters, womanizers. Plenty of alcoholism on my Native side, too."

"Sorry," I mutter lamely, not knowing what to say.

"Yeah, booze killed off my Native side and made my Mexican side crazy," Fernando says, laughing a bit. "I guess I learned to be aggressive because my people were so fucked up. My Potawatomi grandmother though, she tried to shield me from all that. She used to call me her ray of sunshine, her cute little fat boy," Fernando says with a chuckle.

"I feel it's her spirit that's protected me throughout my life. Like when I injured my hand. That machine could have eaten not just my hand — it could have eaten *me*! My grandmother's spirit saved me then and a bunch of other times."

After a bit, Fernando continues. "Growing up, I had no control over my environment, over my family, over substance abuse. One thing that appealed to me about being an ironworker is you have to be hyper-focused. You have to control your environment so you come home in one piece. But that's a problem, too, 'cause now I wanna control my son, my girlfriend, even this March. I think this March makes me angry 'cause I can't do anything about all the goofy shit going on."

We walk in silence for a while, until Fernando asks, "When things like that happen to you, it's gotta mess you up, right?"

"Sure, but not forever," I speculate. "At some point, we've gotta stop blaming our parents, or some bitter old hag of a nun, or some child-abusing swim instructor, or an alcoholic father. If Nelson Mandela can survive thirty years in prison and come out strong enough to lead his people to freedom, the rest of us have it easy. Heck, you've wrestled with some pretty bad shit, too, from the sound of it. You're still an angry sonofabitch, but you've made progress, right?"

"I suppose," Fernando says, thoughtfully. "I came on this March because I want to make a difference. And I think I am, but it's so hard to deal with all these nutty people — Lala with her power trips, Gavain with his insane ideas, and some of these rich kids who just seem like spoiled brats."

As the road narrows, a whole convoy of trucks barrels down the mountain toward us. I say to Fernando as we fall into single file, "Now's a good time to stay hyper-focused, hey?"

"Yeah, you got that right," yells Fernando as the first truck speeds by. When the convoy of nine or ten semis has passed, Fernando continues, "You know, I'm realizing something. A door closed when that machine crushed my hand. But five other doors opened. If I hadn't had that accident, I never would have sought out my Native heritage. I wouldn't have started facing my issues — like why I get angry, why I'm anti-social, why I sabotage my relationships. And I wouldn't be on this March, wouldn't have met you, wouldn't be having this conversation."

"That's pretty powerful, man," I say. "Honestly, I don't have many answers. But I know we can't blame our shortcomings on bad shit that happened in the past. We've gotta be strong. And we need good friends, a solid community, meaningful work. Work like this. What we're doing is hard — but it's important."

We walk in silence until we see other marchers and the break truck on a bend in the road ahead. "For myself, when it comes to that swim instructor, part of me would love to see him in that dunk tank in Hell," I say. "But a higher part of me knows I've gotta forgive him. It's not only the right thing to do, it's where I find my own redemption. That's where I get to heal, to move on, to put the pain of that experience behind me. I guess I've done that without really thinking about it, cause as far as I can tell, that instructor has zero power over me today."

I glance over at Fernando, smile and say, "I even learned how to swim. Eventually. In my late twenties. I suck at it, but at least I'm not afraid to put my head in the water."

Fernando laughs and says, "Well, if we have a chance to go swimming before the end of the March, I promise I won't mess with you when you're in the deep end."

Toward the end of yesterday's 20-mile march, Fernando and I stopped at a bar in Greensburg for a drink and to watch the Pittsburgh Penguins game. We traded small talk with other patrons, but when I told them about the March and our concern about climate change, the conversation quickly moved back to the hockey game.

Today, I walk by myself. I pass two men renovating an old building slated to become a pet boarding center. "Really?" I ask. "People out here in the mountains pay someone to take care of their pets?"

"Yeah, there's all kinds of rich executives in town," one says as he laughs. "They'll drop off their puppy in the morning, drive to a high-paying job in Pittsburgh, then pick up their dog on the way home. But hey, it keeps me employed."

As often happens in man-talk, the conversation rolls to sports. I gain some respect for knowing a bit about last night's Penguins' game. The two men are curious about why I'm walking. When I explain, one shakes his head and says, "Three thousand miles! That's amazing. I couldn't do it."

The man pauses, then continues, "You know, you might have a hard time talking about climate change around here. Nearly everybody's got a friend or a family member with a job in the coal industry."

The next day is long and difficult — 21 miles up and down more steep hills than I can count. I break up the day with lots of conversations. I talk with men trimming trees along the road. During lunch at a pizza parlor I talk with the owner. He's from Mexico and tells me about a long walk friends of his did there. Later, I visit with a man burning leaves and sticks in his yard. Throughout the day, I politely try to insert climate change into the conversation, but it's always a non-starter. The painter I met this morning was right.

We stay at a church again tonight. Most marchers bed down inside, but a few of us set up our tents. After dinner, as I'm heading out the door to my tent, Gavain and Dana's 6-year old daughter, Tilly, hands me a fragrant cinnamon carnation that Mary DeCamp salvaged from a dumpster. I'm moved by such a sweet, innocent gift, rescued from a landfill and given to me by a child. I fall asleep with the flower tucked under my nose, hoping for more moments like this and wishing the emotional roller coaster of life on the March would magically mellow for the remaining miles.

A MEMOIR FROM THE GREAT MARCH FOR CLIMATE ACTION

CHAPTER 41

COAL COUNTRY

"Coming from a coal region, I know that coal is more than making a living, it's a way of life. It's not just your job, it's your community."

— JOE BIDEN

The temperature is 36° when we wake on October 20. I set out on my own today, and walk past homes as rough and weathered as the mountains themselves. Several front yards sport signs that read: "*Stop the war on coal: fire Obama.*" Many of these homes are heated with coal.

I recall Dublin as a kid, where the air was so thick with coal smoke that it clawed at my lungs and burned my eyes. But here, the thin wisp of smoke drifting slowly across my path from a single coal-heated home is pleasant, with an odd sweetness to it.

Walking past the Coal Miners Cafe I'm intrigued by a sign that reads, "*Miners Meal Deal: Eat it all, get it free.*" The challenge features a two-pound burger, fries, coleslaw, and a massive mound of dessert decadence called Gob Cake Sundae.

"Eat it all?" I ask myself out loud. "No problem!" But the Miners Meal Deal doesn't kick in until later in the day, and my vision of glory through gluttony remains unrealized.

I stop anyhow, as I'm greedy for a second breakfast and have phone calls to make. I settle in at a table by the window and place my order. The cafe's walls are heavily decorated with mining memorabilia, much of it from older times — pick axes, coal shovels, kerosene lanterns, portraits of miners, statues of miners, statues of the mules that used to work in the mines.

An hour later, I pay my bill and make sure the server notices the generous tip I leave. Her name is Jody. She's been pleasant and conversational, so I ask, "Can you tell me why this place is called Coal Miners Cafe?"

Jody looks around the restaurant. There are only two other customers, and they're just digging into their breakfast. "Come," she says. "Let me show you something."

We walk toward the front of the restaurant, then veer off into a small room set aside as a memorial to the nine miners rescued in 2002 after surviving three days in a flooded mine shaft. The walls are covered with framed photos. Above them sits a placard that reads, "Believe in Miracles." Indeed, when Jody shares the details of the rescue — how the miners were preparing to die, how rescuers found a way to pump oxygen into the mine shaft, how the miners' arms and legs were purple when they were finally rescued — it does seem like a miracle.

"Miners are our heroes. They're so brave," she says. I nod in agreement, and she continues, "Coal is about more than just jobs. It's how we stay warm, cook our food. It's who we are."

"Kinda like corn is to Iowa," I respond. "We've got special machines just for growing corn. We've got 'corn feeds' in the summer. South Dakota has an enormous corn palace. Oh, and I've even got a friend whose grandfather was the US corn-shucking champion."

We pause to admire one of the photos. The miners' faces, blackened by coal dust, are proud and harsh but not unkind. These are the faces of men who work in conditions unimaginable for most of us. I can't help but see beyond the images to the bigger picture — the high incidence of black lung disease among miners, the fatal occupational injury rate six times greater than in private industry,[1] the destruction of these mountains, and the climate impact of the carbon released when this coal is burned.

"Corn is our culture in Iowa, but there's a lot of problems with it," I confess. "Soil erosion, water quality, chemicals, monoculture. Someday, Iowa's gonna have to move beyond corn. That'll be tough, economically and culturally."

---

[1] https://www.bls.gov/iif/oshwc/osh/os/osar0012.htm

"I hear that," says Jody. "The world's already moving beyond coal. Losing our jobs is bad enough, but losing our heritage is even harder."

"How do people in coal country feel about our group talking about climate change, calling for America to get off fossil fuels?" I ask.

"We've known you were coming for a few days and there's been some chatter about it," says Jody. "Folks don't like your message but they respect what a big challenge it is to walk across the country. People may not agree with you, but they admire your commitment."

She pauses, then adds, "I'll tell you this: there's more and more concern about fracking and what it's doing to the water in some places. There's also worry about mountain-top removal. Miners like their jobs, but when you ask how they feel about a mountain near them having its head blown off, a lot of them aren't crazy about it."

Jody has to get back to her tables. I thank her for taking the time to talk. It's the first heartfelt conversation I've had with someone about coal since leaving Pittsburgh.

Sarah pulls up as I'm preparing to leave the cafe. She hands her keys to a passenger and hops out. "I so need a break from logistics," she tells me. "I'm here to walk!"

"Great! You and I haven't walked together since Arizona," I say. "We're overdue."

A couple miles down the road we pass a sloping field packed with thousands and thousands of wrecked cars and trucks, piled ten high in places. "Whoa, a car graveyard," says Sarah. "I bet most of these vehicles were in a crash."

Looking through the fence, we see where air bags went off and jaws of life were used. I wonder how many dead bodies were pulled from the countless vehicles that have found their way here over the years. On the edge of the lot sits a stately, two-story brick house. Long abandoned, the remnants of soiled curtains hang in cracked windows. It seems appropriate that a decaying home presides over this kingdom of mangled steel and plastic.

"Bet this makes you feel good to be out walking today instead of in a car," I say to Sarah. "You know, in the last eight months, I've only been in a car three times for any kind of distance. Once was with you scouting the route between Scottsdale and Payson, back in Arizona."

"Yeah, that was the first day I left some of my dad's ashes along the route," recalls Sarah.

"I'd forgotten about that!" I say. "That's right. We took a detour on a gravel road and stopped to admire that incredible view of the mountain range. You took a pinch of your father's ashes from a purse and let 'em go into the air."

"I was planning to leave a pinch here and there all along the March route, always at special places," recalls Sarah. "But something really strange happened in New Mexico. We were camped in that sacred canyon on Laguna Pueblo land. You remember how special that canyon felt?"

"I sure do! *Everyone* on the March got along with everyone that night — even with me," I laugh.

"Well, as usual, I was the last one out of camp. As I was packing my gear, I left some ashes there. Then the purse just vanished, how I'll never know. It was so weird. I spent twenty minutes looking for it. I was so upset. Then I had this 'ah-ha' moment and realized that's probably where my Dad wanted to be."

We continue on the road past a huge billboard featuring a larger-than-life, three-dimensional Jesus and the words, "America must end the terror of abortion." Later, we meet a man tending an impressive garden of dahlias. We linger to admire the flowers. The man picks a half-dozen for Sarah. He insists that I have one as well, which Sarah fastens to my lapel.

"There, now you look better," laughs Sarah.

"So, tell me more about your dad," I say.

"When I was a kid, he talked a lot about climate change. He was passionate about it. I've always been concerned about it, but it's never been a huge priority. This March reminded me of the mess we're in, and reconnected me with my dad."

"Wait, so your dad was talking about climate change back in the 1970s?" I ask.

"Yeah, incredible, huh?" says Sarah. "At his memorial service in 2008, the conservation folks who showed up told me back then even *they* thought he was nuts."

I think of how hard Sarah has worked and how thankless some of the marchers have been, so often not appreciating her efforts. "Are you glad you took this job?" I ask.

"Well, yeah, I needed an income," Sarah laughs. "But this is a way for me to honor my dad. If he were alive, he'd be walking with us. Since that wasn't possible, obviously, I brought his ashes with me."

"Funny," I say. "Here we are walking together, just the two of us, yet it feels like a foursome, doesn't it? Through my walking stick, especially when I'm walking alone on a really tough day, I feel that my dad is marching along with me, encouraging me to keep going."

"Same for me," says Sarah. "There've been times when I've wanted to quit. But I'd think of my dad and knew if he were here, he wouldn't give up. He'd be encouraging not just me, but all of us to keep going, to keep spreading the word."

I arrive in camp to more conflict. Gavain and Dana want the gear truck to remain in Washington, DC, for possible use in direct action. They're evasive about details, but eventually admit that the truck could be used to block the entrance to FERC. My curt response is, "Nope! Not a chance. That truck's gotta haul the solar collector, wind turbine, and a whole lot more back to Des Moines. I'm not gonna risk having it end up in that salvage yard we passed today." Dana calls me a dictator.

Sarah has a hard time finding a campsite in Breezewood. The pastor of one of the town's only churches wants to accommodate us, but his congregation refuses. In an act that impresses Sarah and me as distinctly Christian, the pastor explains, "I asked myself, 'What would Jesus do?' and the clear answer was, he'd put you up regardless of what the congregation says."

He turns over his small ranch home to us for the night. With so many marchers wanting showers, several of us wait nearly two hours for the hot water tank to refill. It's a festive environment.

Michael cooks tacos in the pastor's kitchen. Tired bodies sprawl on the couch, on every chair, and all over the carpet. It's a challenge just to walk through the house, which has the feel of a 1960s commune minus the drugs and sex. I have to wonder if the pastor will still have a job tomorrow.

About two-thirds of the marchers sleep inside while the rest of us pitch our tents on a ridge overlooking a noisy colony of chain restaurants, hotels, and service stations catering to travelers on Interstate 70. During the night, most highway sounds are drowned out by a strong wind that howls through camp. We haven't experienced a wind like this since the high deserts of New Mexico and Arizona. I'm grateful my tent is sturdy enough to withstand the wind's force.

We set out the next morning as local supporters warn us of a recent march forced to quit due to the difficulties of the next stretch of road. I tell them I'm not worried. We're now so close to our destination that some of us would drag ourselves to the White House if we had to.

Today's elevation changes are indeed a challenge. I understand our local supporters' concern. Some of the drop-offs along the road are extreme, and I hope marchers are cautious. I stop frequently to catch my breath and admire the view. Walnuts lay thick along the roadside in several places. Curious as to how far one can hit a walnut given this uncustomary advantage of height, I use my stick as a bat to send one after another flying off the mountain. Never before have I hit round objects so far. Walnuts sail off the steep slope, discoloring my stick with a gooey, rich, yellow-brown stain. When I'm done playing home-run derby, I try with minimal success to remove the stain with my blue handkerchief, which I suspect will remain forever tinted with tones of yellow and brown.

At the bottom of one particularly long, challenging stretch of mountain road I stop at a quaint little shop that offers a limited selection of beverages, junk food, and eggs. Alongside these items are clothes, books, and some antiques. The store feels practically dead. I browse for a minute until an older woman enters through a back door. She smiles and asks, "How can I help you, sir?"

"I'll have a bottle of water," I say, even though my bottle is full and bottled water is a scam. "Is it ok if I sit on the bench out front and make a few phone calls?" I ask.

"Of course," she replies as she rings up the bottle then exits through the back door, trusting me not to plunder her meager inventory.

During the forty-five minutes I sit out front not a single customer stops by. I feel sorry for the woman. I wonder how she survives. I bristle a bit as I consider how government uses her taxes to subsidize the malls, big box stores, and chain businesses determined to put her and other small operators out of business.

I think about the contrast between her ghost of a shop and the bustling chains we walked by this morning. She's not a victim of the "free market." She's a victim of corporate socialism, of a bi-partisan oligarchy that uses the power of government to wage a relentless war of domestic colonialism. The oligarchy centralizes economic power as it seeks to crush independent doers and thinkers. It despises all that is small, humble, and defiant. It rewards that which is large, brash, and compliant. At the root of the oligarchy's lust is an addiction to money. As with any addict, it constantly craves more and more.

It's no secret that the oligarchy has done very well for itself. Today, the top one percent of Americans earn three times as much as they did in the 1980s, while the bottom 50 percent have seen no growth in earnings.[2]

The door opens as the owner slips quietly back into the store. She putters around for a bit, perhaps curious about why I'm still sitting out front long after I've finished chattering on my phone. But she doesn't seem inclined to talk and I'm not in the mood for conversation either.

So I pack up my satchel, thank her once again, and set out singing a Greg Brown tune, "Your Town Now":

> *"From the mountains to the plains*
> *All the towns are wrapped in chains,*
> *And the little that the law allows,*
> *And it's your town now, it's your town now....*
> *Don't let 'em take the whole damn deal,*
> *Don't give up on what you really feel.*
> *Ah, the small and local must survive somehow,*
> *If it's gonna be your town now...."*

---

[2] http://money.cnn.com/2016/12/22/news/economy/us-inequality-worse/index.html

I think about all I've seen walking across this country — this great, vast, diverse, amazing country. I think of all the little cafes I've eaten at, the many old hardware stores I've shopped at for stick tips, and the countless shuttered store fronts I've walked past on struggling Main Streets. When I compare the thriving blight of Breezewood's chain stores with the languishing charm of the shop I just left, it's clear that things are stacked against Main Street, against entrepreneurs, against independence.

Thomas Jefferson said, "I hope we shall . . . crush in its birth the aristocracy of our monied corporations which dare already to challenge our government to a trial of strength, and to bid defiance to the laws of their country."[3]

I'm sorry, Mr. Jefferson, but we've failed, and failed badly. Instead of crushing the monied aristocracy, we've let it crush us, let it drown freedom in a swamp of greed and consolidation. When — or in all honesty, *if* — we solve our climate problem, the equally intractable challenge of toppling the oligarchy awaits.

---

[3] http://tjrs.monticello.org/letter/1392

A MEMOIR FROM THE GREAT MARCH FOR CLIMATE ACTION

CHAPTER 42

I LEAVE THE MARCH

"Nomads travel light and have a minimalist mindset. It means that they consume experiences instead of accumulating Stuff."

— FROM BECOMENOMAD.COM

We cross into Maryland today — our final state before arriving in Washington, DC. A few locals join us, including an astronomer who finds planets for a living. I joke with him and ask if he gets paid on commission.

Bruce Nayowith, a psychiatrist from western Massachusetts, walks with me for a few miles, too. He's aware of the March's internal conflicts and the hostility toward me. He suggests quite seriously that the Oedipus complex is partly to blame — where psychologically immature people rebel against the perceived father figure.

"Yeah, that sounds about right," I concur.

I hear murmurings this morning of another absurd proposal: canceling the rally. A general camp meeting is planned tonight and Gavain is slated to facilitate. A mile or so from our destination at a church north of Hagerstown, I ask Gavain if canceling the rally is on the agenda. He assures me it isn't.

Once at camp, I set up my tent with other tents clustered tightly in the small churchyard. Before answering email and making calls, I attend the meeting to brief marchers on staff activities. I sit next to Gavain and ask him again, point blank, if canceling the rally will come up for discussion. He again assures me it won't.

I leave the meeting and spend the next hour working. It's a beautiful evening. I set up my "desk" on the church's patio, a stone's throw from where the meeting continues. After it adjourns, we enjoy dinner prepared by our hosts. As I'm finishing my meal, Kathe approaches and says, "Well, you probably oughta know that someone brought up canceling the rally at the meeting, and it passed."

"What?" I shout in disbelief, practically spitting out a bite of food. "Twice this afternoon, I confronted Gavain. He told me there would be no talk at tonight's meeting about canceling the rally."

"Well," said Kathe, "someone brought it up after you left."

Kathe won't say who brought it up, but I suspect Lala. "You don't just bring something up without proper notice," I say. "We've been working on this rally for months. You can't just cancel it with a week to go."

"Well, we voted on it and agreed not to do it," says Kathe.

"How many people voted to do this?" I ask.

"Nearly everybody," Kathe says.

"You're shitting me," I say in genuine shock. "And you? How did you vote?"

"I voted for it, too," Kathe says.

Now I'm red with rage. I swear at Kathe, dropping a string of f-bombs. "So not one damn marcher, not even you, had the decency to come get me when this was brought up. I'm working my ass off to keep this March going and you couldn't even let me know? Well, fuck you and fuck everybody," I yell as I storm off.

I walk away from the church and into a nearby park where I find a swing attached to a high branch of an old chestnut tree. I sit there for close to an hour, swaying back and forth, trying to calm down, watching the sun set in the west. I'm so angry, possibly angrier than I've ever been in my life, that it takes a long time for me to settle down, to think rationally, to sort out what to do.

I'm disgusted with marchers — not just with Gavain and Lala, but with those I usually get along with, like Kathe. I feel betrayed and disregarded. It's as if stupidity, deceit, and rebellion have become the guiding principles of the March.

I'm disgusted with myself, too. Kathe's complicity in what Gavain did was wrong. But I didn't have to explode like that. My reaction feels so out of character. In my life in Des Moines, that almost never happens.

What to do? What to do? In the distance, I see a lone whitetail buck, silhouetted against the darkening sky, walking by himself. I watch him for a minute or two. Once in a while, the buck bends his neck to the ground to eat. Mostly, he walks slowly and deliberately, heading south with the fading colors of dusk behind him.

Suddenly, I see my way. I'll leave the March and depart immediately! I crave peace after being treated so badly, not just tonight, but regularly over the past seven months. I crave peace, but I also deserve exile — exile for having acted so inappropriately, on so many occasions, and in a manner inconsistent with my commitment to nonviolence. The reward of peace; the punishment of exile. If I leave, I achieve both.

I feel a rush of exhilaration, as if I've at last found the courage to leave an abusive relationship and take the plunge into a frightening yet liberating unknown.

I'll walk the next 85 miles on my own and meet up with the March on November 1 for the final seven miles. Rally or no rally, when we get to Lafayette Square across from the White House, I'll stand on a bench or a stump or some makeshift soap box and give my speech, even if only birds and squirrels are present to hear it.

I walk back to camp. Everyone is either in their tent or already asleep. I quietly pack my bags, take down my tent, and carry them to the gear truck. Doug Cooley is the lone marcher at the gear truck, and I tell him my plan. Doug used to work as a psychotherapist before retiring to travel the world on his sail boat. He knows the conflicts I've been dealing with, and understands why I'd want to leave. I ask him not to let others know I've left until tomorrow afternoon. I load the bare essentials into my satchel, tie my raincoat around my waist, pick up my stick, and walk out into the night.

I arrive in Hagerstown around 10:00 p.m., hoping it's not too late to find a hotel room. On the way into town I study the options on my phone. The first accommodation listed is the Dagmar Hotel in downtown Hagerstown.

Arriving at the Dagmar, I walk by three young men standing in the entrance. Their faces are rough, hewn with lines carved by frequent scowls and infrequent smiles. They barely move to let me enter. Inside, a pleasant Indian man with an English accent greets me and shows me to the front desk. He says they have a room available for $50. I ask if I can see the room, and a man named Ricky leads me to the elevator.

We ride the elevator to the fifth floor where a young woman meets us in the hallway. She smiles and says, "Ricky, Ricky, please give me a dollar. I need to buy a slice of pizza." The woman is practically slobbering. She paws at Ricky as we walk down the hall. Ricky opens the door to my room. When we enter the woman comes in with us.

"Please, Ricky, just a dollar, it's all I need," she insists. "Just one dollar so I can get some pizza." Then her eyes get wide, and she looks right at Ricky and says, "I've been really good you know, and I've been trying to help. Today, I killed a whole bunch of cockroaches!"

I look at the room and notice the thin, swayed twin mattress — and nothing more. That's it. There's nothing else in the room, except for a lone shade on the window, which is broken. I pan the room for cockroaches as Ricky tells me the bathroom is down the hall. The carpet appears to have been installed when wall-to-wall carpet was first invented. I imagine some of the stains are the decayed bodies of squashed cockroaches.

I'm exhausted and have no interest in walking any further. This may be the only hotel room left in town. But if I stay here, I won't sleep and may well leave in the morning with one or more insect species as traveling companions. I thank Ricky and tell him I'll pass. On the way back to the elevator, the woman continues her relentless pleas for pizza money and her braggadocio about the slaughter of cockroaches.

I slip past the three scowling sentinels and back onto the street, hoping I don't end up back at the Dagmar as my only option. I walk another mile and find a room at a hotel on the eastern edge of town. It's respectable, costs $60, and I don't have to share it with cockroaches.

A MEMOIR FROM THE GREAT MARCH FOR CLIMATE ACTION

On October 27, after a meager breakfast and an insipid cup of tea, I head out in the general direction of Washington, DC. It's both freeing and unnerving to have no planned route. I track generally southeast, avoiding the route of the March and seeking quiet roads.

On the outskirts of Hagerstown, I pass three road workers. "I see we're all wearing the same outfit today," I say, referencing my yellow safety vest.

"You reporting for work?" jokes one of the men.

I pretend to grab a shovel. "Tempting," I say, "but I hope to meet with some politicians in Washington next week, so I'd better keep moving."

"Why bother? That place is a mess," another man says.

"I know," I reply. "I'm gonna try to straighten it out."

"Good luck," he says, "but I think it's beyond fixing."

Today's walk is full of green fields, warmth, and cattle. I revel in the smell of cow life — the mix of silage, fresh-cut hay, manure, the cows themselves. Unlike the cattle I saw in Colorado's feedlots, where thousands of miserable beasts stood crammed into grassless pens of mud and feces, these cows enjoy a comparatively idyllic life. I admire how peaceful they are, oblivious to the problems of the world, oblivious to their own inevitable demise, living entirely in the moment and content to have their basic needs met. If cows could work themselves into the lotus position, they would be Zen masters in no time at all.

"This is my week to be a cow, ok?" I say aloud to a young Hereford steer who slowly chews a mouthful of grass as he watches me pass. By mid-morning though, I confront the reality that I can't simply lie down in a pasture tonight and sleep with the cows. I call Bob Warner, who helped organize our border crossing yesterday, and ask if he knows anyone in the area who might provide me a bed and a shower. He puts me in touch with Monica and Roy Greene, who are happy to accommodate me.

I stop for lunch at a restaurant in Boonsboro. Abandoning the carefree life of my Hereford friend, I spend the next two hours thanking recent donors and making phone calls. Tom Vilsack has offered to help line up a meeting for marchers with President Obama, or if that's not possible, with the President's key climate staff. I've been playing phone tag with Allison Hunn, the White House scheduler. Allison and

I finally connect today. It's still not clear if the President or his staff will have an opening while we're in DC, but at least we're now on the White House's radar.

By late afternoon, I've covered eighteen miles and arrive in Middletown. Monica meets me and drives me to her home. After a hot shower, I join her and Roy for supper. Roy is a career Army medical personnel; Monica, a passionate climate activist. She's one of the calmest, gentlest, and most present people I've ever met.

After an excellent night's sleep, Monica prepares a breakfast of scrambled eggs, cherry tomatoes, steel-cut oatmeal with walnuts, cinnamon, and raisins. My cup of tea is as excellent as yesterday's tea was dismal.

Monica drops me where I left off yesterday. My plan is to continue southeast to Point of Rocks, then follow the C&O Canal Trail along the Potomac River for the final 50-mile stretch into Washington. A scenic, traffic-free trail along one of America's most famous rivers — what a perfect way to end the March! Yesterday and today, I've pretty much had the roads to myself. Even better, for the final three days, it'll be just me and an occasional walker or bicyclist.

During the last few miles to Point of Rocks, where Monica will pick me up tonight, I consider what it will be like for me and other marchers to return to our former lives. Nomadism is so deeply ingrained in the human genetic code that it didn't take marchers long to adapt to the fluid life of the March.

But what about the readjustment to modern living? Sure, there's the appeal of warmth, dryness, ample food, a consistent water source, and legs and feet that aren't constantly sore. But when one has grown accustomed to moving at three miles an hour instead of sixty, to living communally instead of alone, to being outdoors instead of in buildings, to confronting the breathtaking rawness of an ever-changing landscape day after day — is one forever spoiled for "normal" living?

I think about what passes for normal — sitting in cubicles, staring at screens, eating food made in factories, conveying ourselves without ever moving our legs. Is there anything about such a life even remotely consistent with how humans are wired?

Natural or not, that *is* our life, our world, our community. It's the paradigm within which we must live, work, love, and struggle — a paradigm that comes with many blessings even though it may be antithetical to our basic identity.

I arrive at Point of Rocks where Monica's waiting for me. I'm grateful tonight for a bed, a shower, and a hot meal. Yet as I slip into the car, I feel a twinge of the fight-or-flight mechanism, like a wild animal entering a cage. The moment passes, and I'm comforted knowing that I have no idea what tomorrow will bring.

CHAPTER 43

PILGRIM

"Real love is a pilgrimage. It happens when there is no strategy, but it is very rare because most people are strategists."

— ANITA BROOKNER

I work at the Greene's the next morning while I have a decent Internet connection. Since Chicago, I keep hoping the demands of the March will taper off so I can walk without distractions, but that hasn't happened. Even after leaving the March my workload shows no sign of letting up. Donations at our overnight stops have been hit or miss, so I continue to raise money. Logistical challenges for our DC march and rally need to be addressed. I'm getting two or three press enquiries a day and the documentary crew wants an hour of my time every other day. Most significantly, the push for a meeting at the White House is at a critical stage.

After a productive morning of work, Monica drops me at Point of Rocks. It's a perfect day for walking, and I set out late morning on the C&O Canal Trail. Fifty miles separate me from the end of the March. I had hoped to knock off 18-20 miles today, but the only lodging I'm able to locate is at a hotel across the Potomac River in Leesburg, Virginia. So I opt for a 13-mile walk and resign myself to a longer day tomorrow.

The trail is practically people-free, though I share it with all kinds of birds and mammals. Unlike the flattened creatures I've seen along America's highways, these are still alive. On a bend in the trail, three deer emerge from the trees seventy yards in front of me. One of them is a yearling. All seem small compared with Iowa's corn-fed whitetails. As I approach, the two adults bolt into the forest with the yearling clumsily in pursuit.

A bit farther, I hear a loud splash. Curious, I cut quickly through the brush to see a six-point buck crossing the Potomac. At first, he walks, then swims as the water gets deeper until all that remains visible is his head. Given the current and the buck's inadequate design as a swimmer, I wonder what it is about the far shore that compels him to attempt such a difficult trip. Perhaps, if the deer knew I had walked here from the West Coast, he'd wonder the same thing about me.

Later, I come across a groundhog whose autumn plumpness makes its sudden escape difficult and comical. I see more squirrels than I can count, and the bird life along the trail is abundant and varied. I lose myself in the beauty of my surroundings and the walk goes by quickly.

My phone rings twice today. First, it's Steve telling me he's going to join me for the final two days of the March. He's rented a hotel room in Bethesda, and I'm welcome to join him there on October 30 and 31.

Then Sarah calls to tell me Fernando was summoned before the judicial panel yesterday and thrown off the March. I'm sad to hear this but not overly surprised. Sarah says the final straw was when Fernando exploded yesterday at Kim and Liz after he found they had thrown his new $200 jacket on the ground because it was in the wrong place in the gear truck.

"That makes two March exiles," I say to Sarah. "One of us by choice, one by judicial fiat." Sarah also tells me that marchers had another meeting and decided to do the rally after all. "Nothing surprises me anymore," I say, shaking my head.

Late that afternoon, I arrive at White's Ferry, which will carry me across the Potomac River to catch the shuttle to Leesburg. Cars pay $5, but as a pedestrian my charge is only $1. I watch the ferry approach from the Virginia shore. As it gets closer, I notice garish Halloween creatures decorating the boat — a bloody pig's head with a snake in its mouth, a rotting zombie strapped to the mast, and various disgruntled skeletons strewn about the deck. It's a thoroughly tasteless display that screams, "There's nothing politically correct about this ferry and if you don't like it you can take the long way around."

I search for "White's Ferry" on my phone and learn of its colorful history. The ferry began operations in 1786. At one time, 100 ferries worked the Potomac. White's is now the only one.

The boat is named after Confederate General Jubal A. Early. Like Early, owner Edwin Brown is a rebel who goes well beyond adorning his boat in a manner sensitive palates might find offensive. For four years, Brown defied US Coast Guard orders to cease piloting the Early with an unlicensed mariner.

As I watch the thick cable drag our boat through shallow water across the Potomac, it is indeed hard to understand why a mariner's license would be needed. But Coast Guard officials were adamant. "Frankly," claimed Deputy Commander Jonathan Burton, "everything is on the table, including civil penalties and criminal penalties."[1] Burton had served in the Coast Guard for twenty years and said he had never encountered a more flagrant example of someone flouting the law.

Brown responded defiantly, "It'll be a cold day in hell before they collect any money from me."[2]

Collect any money? What an understatement! The Coast Guard threatened fines of up to $32,500 *per trip*. With multiple trips every hour over an 18-hour day, that could add up to over a million dollars *per day*.

Incredibly, Edwin Brown won this fight against a transparently ridiculous regulation. He continues to operate White's Ferry with an unlicensed mariner.

As the waters of the Potomac churn underneath, I think back to the little shop on the mountain road east of Breezewood. "I'd like to put Mr. Brown in touch with the woman who owns that store," I say to the zombie hanging on the mast. "Big Government and Big Business. What a team, hey? Here you are, just hanging out on this boat, trying to earn a living, and all they can think about is making life miserable for average zombies like you."

I pause and pretend to wait for the zombie's response, wondering what the passengers sitting in their cars think of a guy with a walking stick talking to a creepy Halloween ornament. "Yeah, I get it," I say to the zombie. "You don't give a crap. You just want a chunk of my brains. Well, sorry, none to spare."

---

[1] Mutinous Ferry Roils the Waters, Washington Post, September 15, 2006

[2] Mutinous Ferry Roils the Waters, Washington Post, September 15, 2006

A MEMOIR FROM THE GREAT MARCH FOR CLIMATE ACTION

My intention to land in bed early is thwarted by three hurdles — two of my own doing, one foisted upon me. The documentary crew wants to interview Miriam and me together. So the videographers pick up Miriam at the March camp and drive over to meet me at the hotel in Leesburg. It's my third interview with them in as many days.

After the interview, I convince myself it's my duty as a Midwesterner to watch the Kansas City Royals in the final game of the World Series. After leaving the bar, I discover a baby grand piano in the hotel lobby. I play for a full hour, grimacing each time I forget a passage of one of my favorite Chopin nocturnes. I've played piano just a handful of hours these past eight months and worry that this extended hiatus will leave some of my repertoire beyond salvageable.

When I finally crawl into bed after midnight I'm wide awake from so much activity. I think back to a night in September when several marchers had home stays. My host was a woman named Rachel, who was passionate about our cause. When I arrived at Rachel's home, the smell of herbs and tenderly prepared meat and vegetables greeted me at the door.

We dined outside in the garden under a canopy of trees interspersed with carefully tended beds of flowers and vegetables. The food and wine were exceptional, the conversation even better. I shared with Rachel a quote from Epicurus: "Before you eat or drink anything, consider carefully who you eat or drink with rather than what you eat or drink, for feeding without a friend is the life of a lion or a wolf."

Rachel appreciated the quote. After dinner, she suggested a walk. We strolled over to a nearby park, then followed a path that ran along a creek. We paused to laugh at the antics of some ducks, and at barefoot college students practicing the tango.

Leaving the park, we passed a cafe. I offered to treat us to dessert and another glass of wine. We sat outside, enjoying the beautiful evening, talking for another hour.

Back at the house, Rachel offered me a foot massage and I hungrily accepted. Her hands worked gently into my sore feet. I closed my eyes and couldn't imagine anything in the world that could feel more perfect.

We talked a while longer until I realized it was 11:00. I didn't want the night to end, but told Rachel I had to get to bed. In what felt like an absolute perfect end to the evening, we stood up, hugged, and said goodnight.

I woke at 5:30 the next morning and quietly prepared to set out. Marchers were assembling at 6:00 a half mile away and I needed to be there on time. I didn't want to wake Rachel but couldn't leave without saying goodbye. Her bedroom door was open, so I walked over to the side of her bed, bent over, and whispered, "Hey, I'm leaving now. Thank you for everything."

I couldn't tell whether Rachel woke up or was still mostly asleep, but she smiled as I stood there trying to sort out the appropriate degree of affection for departing under such circumstances. I reached out my hand and clasped hers, thumb to thumb, the way two close male friends often greet each other. This felt awkward, so I shifted to a standard handshake, holding her hand for a while, relishing its softness and warmth. I didn't want to let go but had to change positions, so I moved my hand into a finger-clasp. Rachel smiled again, and as I said goodbye I still couldn't tell if she was mostly awake or asleep. Either way, I was certain she thought my parting gesture the most unusual anyone had ever offered.

I walked out of the house into the pre-dawn darkness and smiled at how wonderful our short time together had been. For a handful of hours, I'd felt like Rachel and I were a couple, not sexually intimate, but enjoying a deep intimacy of mind and spirit, of food and drink, of movement and stillness, of eye contact and touch.

Lying in my hotel bed in Virginia, tired but wide awake, I'm overcome by a sudden sense of clarity. There is no Dulcinea! Victorian Man is a myth. Waiting for an imaginary soul mate is an excuse to avoid the hard work of building a loving relationship, of addressing the inevitable conflicts, of learning to work at compromise and sacrifice. What Victorian Man truly desires is to meet the woman who is as perfect as he believes he is. Victorian Man wants nothing less than to fall in love with the female version of himself. Victorian Man is the ultimate narcissist.

That's what my dreams have been trying to tell me! That's what my night at Rachel's confirmed, though I didn't understand at the time. I think back to the statue of the pilgrim of Chimayo. The pilgrim knows where he's going even though the journey itself is a mystery. He has a goal but no strategy. The pilgrim puts his faith in a higher power who he trusts will guide his steps.

Is the search for love all that different from a spiritual pilgrimage? There is no formula, no handbook, no strategy, no predetermined partner waiting in the wings of time. There is no Dulcinea. She's a decoy — a beautiful, alluring, flesh-and-blood decoy — distracting me from the truth that the world is rich with women, one with whom I could share love and happiness, providing I'm prepared to work at it.

Last year, as people signed up to join the March, I told them the experience would change their lives. I knew it would change mine, too, but I didn't expect this. Now, thirty-seven miles from the journey's end, I feel I've grasped one of life's mysteries, one that has eluded me for so many years.

I smile contentedly at this realization. I pull the blankets up to my chin and think of sleep. Yes, there is no Dulcinea. There is no soul mate. I'm sure of it — or at least reasonably sure.

I lie there for a while, a long while, unable to sleep, pleased with what I think I've learned but disturbed that I still can't get Grace off my mind.

CHAPTER 44

BROKEN

"Life is not a matter of holding good cards, but sometimes playing a poor hand well."

— JACK LONDON, *TO BUILD A FIRE*

Since Nebraska, my feet have hurt almost constantly during the day. Sometimes the discomfort is minimal. At other times, it's bad enough to slow me down considerably, though I've never again felt the level of pain I experienced in Cleveland. The adjustments I made that day helped immensely. I'd feared our 130-mile week through the Appalachian Mountains might do me in. Yet my feet handled that challenge surprisingly well, even as shorter days were sometimes a problem.

After a restless night's sleep, I wake this morning to an already bright sky. I glance at my phone. Eight o'clock! For a man with chronic foot trouble and a 20-mile walk ahead of him, sleeping until 8:00 is an invitation to trouble.

Half awake, I scramble out of bed, completely forgetting about plantar fasciitis. When my feet hit the carpet, my legs immediately buckle and I fall to the floor. I grab a chair, then the wall, and slowly claw my way around the room until the pain in my feet lessens, my legs adjust, and I'm able to stand.

Hobbling to the bathroom, I assure myself this is no big deal. I've walked nearly halfway across the country with plantar fasciitis. Certainly, my feet can hold out for another 37 miles.

A MEMOIR FROM THE GREAT MARCH FOR CLIMATE ACTION

I take my time getting ready, then gingerly make my way to a nearby restaurant where I find a table by the window. Without looking at the menu, I ask the server for the restaurant's biggest breakfast, plus two large pancakes. Despite this thrice daily practice of devouring calories and protein like a man fending off starvation, I'm still 24 pounds lighter than when I set out from Los Angeles.

After breakfast, I spend an hour calling local friends and acquaintances in a last-ditch effort to boost turnout for the rally. I'm so wrapped up in work that I forget to pack a lunch. I don't realize my mistake until I'm on the shuttle heading to White's Ferry.

The documentary crew insists on another interview. They meet me on the Maryland side of the Potomac as I step off the ferry. I don't start walking until noon — three hours after having finished breakfast and way too late for a twenty-mile day.

The beauty of the trail helps me forget about both foot discomfort and the climate crisis. The pathway's soft surface eases my feet a bit. The day is calm and the temperature cool with moments of mist or light rain. Between drizzles, sunlight slants through majestic 200-year-old trees. The wildlife along the trail is again abundant.

After three miles, my stomach chastises me for failing to bring a lunch. So far, I've seen no signs of a restaurant or convenience store. I nibble on wild chives, violet leaves, and chickweed. These provide some nutrient value but do nothing to alleviate my growing hunger.

Ten miles into the day, with still no sign of a shop or restaurant, I run out of water. Surely, there has to be a store up ahead. I pull out my phone to look at a map. The battery reads one percent. As I stare at the screen the phone rings. It's Allison Hunn, the White House scheduler. She tells me that a meeting with President Obama's climate advisers isn't possible during the two days we'll be in Washington, but she's able to arrange a meeting in January. We agree on a date, and as I'm saying goodbye, the phone powers off.

For a few seconds, all I can think about is how fortuitous it was that Allison called just before my phone died, and how pleased I am that marchers will get to share our story with the President's closest advisers on climate.

Then the immediate reality sets in — I have no means of navigation, no idea what lies ahead, no way to contact anyone in case of an emergency, no food, and no water. On top of that, despite the soft pathway, my feet are growing sorer with each mile. The poor night's sleep has caught up with me, too. The trail remains almost devoid of people. I have no choice but to keep walking until I find a road that, I hope, leads to food and water.

The sun sets and the temperature drops. Hunger, thirst, fatigue, and the intense pain in my feet drain what little energy I have left. Mysteriously, my growing discomfort magnifies the night's beauty. The river's current is more pronounced, more soothing. The shadowy shapes of the ancient trees guarding the trail come alive, bearing witness to centuries of human and natural drama — both mundane and historic, scripted and unrehearsed. The trees and the river whisper to me to persevere, to fulfill my part in the grand, timeless performance unfolding along the Potomac's life-giving path.

The darkness deepens. When the clouds cover the moon, I can't see the trail. I dig into my satchel for my headlamp and find nothing. I must have forgotten to bring it when I left camp days ago. Walking at a snail's pace, I sometimes stumble over rocks and roots. More than once, my stick saves me from falling.

I pass between rapids on the right and towering cliffs on the left. The last frogs of autumn sing as the half-moon pokes through a thin veil of clouds, reflecting off the Potomac, providing a sliver of light to guide me along the path. Above the cliffs, I hear the sound of a harp and piano coming from a home ablaze with decorative lights that belie a reluctance to relinquish the fading warmth of fall. I stop to listen. I'm captivated, and think of the many hours Kristin and I shared music. I wonder if, perhaps, our marriage might have lasted had we continued to play together.

Fog forms and a light rain starts to fall as I come to a road leading away from the trail. At last, a possible escape route! The road is narrow, unlit, and very steep. I climb painfully up the hill using short, halting strides, digging in deeply with my walking stick, regretting that I have to leave the soft, flat surface of the trail behind.

No food. No phone. No water. But still, I have my walking stick, made from wood that connects me to my father, to monks immersed in lives of prayer, to a forest as venerable as the one I now walk through.

No food. No phone. No water. Yet I feel transformed, as if I don't walk in physical form but as a disembodied spiritual being. My walking stick is no longer simply a piece of wood. It's now Staff, a tool of great power and purpose, a weapon even, protecting me in this battle against fatigue, gravity, and darkness.

No food. No phone. No water. Staff and I are one. Through our union I feel a deepening connection to the reservoir of spiritual power beyond the grasp of normal human thought. The power moves through Staff and pulls me forward.

After a long climb, the road intersects a four-lane highway. At a break in traffic, we cross the highway and head east. There's very little shoulder, then no shoulder at all. We walk half a mile with headlights charging toward us like the fiery eyes of stampeding bison. It's dangerous, even deadly, to walk this road at night in fog and rain, to walk exhausted and with no light. But what choice do we have? I hold Staff in my right hand as protection from the onslaught of raging vehicles, a thin but powerful buffer against injury or death.

Now a bike path parallels the road, and I'm relieved for the increased distance between us and traffic. But after a short distance, the path ends abruptly and we are thrust back onto the highway. The psychological impact is sharp and instant. I feel emotionally and physically crushed and defeated. The spiritual power present just moments ago vanishes. Into the void rushes the cold, numbing nothingness of reality.

I collapse under a stand of trees next to the highway. I know I've been beat, know I can't go on. My March is done, finished. I want so badly to continue, but my body refuses. Knowing how close I came, I'm practically in tears. At least Steve made it. Hopefully Miriam, Jeffrey, and Mack will make it as well. For me, I don't know what happens next. Perhaps if I can get a ride tonight to the hotel where Steve is staying and rest tomorrow, Sarah can drive me to the rally on Saturday.

I fall asleep and dream of a half-naked fairy princess riding a leaf to the ground like a silent green glider. After a few minutes or an eternity — I can't tell — a truck roars by, wailing on its horn, transporting me back to my place under these trees dripping with mist, coldness, and irony.

I don't want to move. I'm not even sure I can. But I'm wet and shivering. I have to at least try to get to where I can find a ride to the hotel. I grope around for Staff, and feel nothing. I panic. I roll to my right. Nothing! I roll to the left. Staff is gone! My panic deepens.

Then I see Staff in the faint light of a streetlamp, lying ten feet away in the grass near the highway. I crawl over, grab it tightly, and let it pull me to my feet. We head east, one slow, deliberate step at a time. I again feel a resurgence of power as Staff leads me forward despite the numbness in my feet, legs, and arms.

We come to a school. The parking lot is empty. I look around for an outdoor outlet to charge my phone, but find none. We continue walking. A block later, we come to a church. There are cars in the parking lot and lights on inside the building. By the doorbell, a sign reads, "Please ring." I press the button. We wait, but there's no answer. I check the door. It's unlocked, so we enter. In the distance I hear a choir. It sounds far away, very far away, as if the church stretches on for blocks.

Immediately beyond the entrance is a lounge. It's empty. We walk in and find a sink. I hastily fill my water bottle, drink the entire quart, then fill the bottle again. Next to the sink, I see an outlet that powers a coffee pot. I unplug it and plug in my phone. Quickly, I pull up a map and see restaurants only four-tenths of a mile down the road. I plug the coffee pot back in and slip quietly back outside, grateful at having avoided detection.

We move slowly along the highway and eventually arrive at a cluster of businesses. I walk up to the first restaurant and tug at the door. Closed! I walk over to the next restaurant. Also closed. A third establishment is closed as well. I feel desperate now, and yank forcefully at the door of the last restaurant, as if pulling the door hard will improve the restaurant's chances of being open.

It *is* open! The host greets me and assures me they're still serving food. I wolf down an entire basket of bread while waiting impatiently for a large plate of spaghetti. I try to eat politely but can't help devouring everything like a wild beast. I ask for a second basket of bread, relieved that the restaurant has few patrons to witness my gluttony.

When my phone is charged, I study the map and learn that Staff and I have walked more than 23 miles today. I let Steve know that we're on our way. I pay $20 for the seven-mile cab ride to Bethesda. Arriving at the hotel, I find my way to the room, thank Steve, and fall asleep the moment my head hits the pillow.

CHAPTER 45
THE WHITE HOUSE

"Here is the test to find whether your mission on Earth is finished: if you're alive, it isn't.

— RICHARD BACH,
ILLUSIONS: THE ADVENTURES OF A RELUCTANT MESSIAH

I sleep late and wake feeling sore but rested. After yesterday's pathetic entry into consciousness — literally falling out of bed — I cautiously set one foot, then the other, on the floor. To my surprise standing isn't a problem.

After breakfast, Steve and I take a cab to the restaurant where I'd ended last night's ordeal. We pay the driver and set out for the eight-mile walk back to our hotel in Bethesda. The walk goes by quickly, even at a leisurely pace, as Steve and I have a lot to catch up on.

I call Fernando and suggest that he join us for breakfast tomorrow before we march to the White House. I invite Doug, too, who has grown increasingly weary of the drama of life in One Earth Village.

On November 1, in a monumental moment of male reminiscing, the four of us banter and trade war stories over large plates of breakfast at a pancake house in Bethesda. Steve opts to meet us at the rally across from the White House while Fernando, Doug, and I head over to Elm Street Urban Park where people are gathering for the final leg of the March.

At the park, a handful of marchers welcome me back. A few greet me with a hug. Two are upset that Fernando has come. I tell them it'd be great if we all just found a way to get along today.

About 150 people set out for the final seven miles of our 3,100-mile odyssey — a group one-tenth the size of our launch in Los Angeles eight months earlier. I'm saddened at how small in number we are, but I'm not surprised. It's been clear for the past two months that our finale would never match the size or enthusiasm of our kick-off.

As we set out from the park, a young anarchist joins us. He waves a black flag and totes a bag of emergency medical supplies. "What's up with the bandages and stuff?" I ask.

"We'll need them if the cops get ugly," he tells me.

"That won't happen," I assure him. "This isn't that kind of march. We've got a permit, and the police are here just to help with traffic. As far as we're concerned, it'll be a boring day for DC law enforcement," I say, both informing and warning the young man at the same time.

He nods, then begins to shout through his bullhorn, demanding an end to capitalism and blaming the free market for climate change. I bite my tongue and resist the temptation to berate the man for hijacking our event.

Doug and I walk together. He tells me he'd come close to quitting a couple times because of the March's discord. "That night you left camp to walk into Hagerstown, I was tempted to join you," he jokes.

"Ha! That doesn't surprise me," I laugh. "So, why *did* you stick around?"

"Well, I'm relieved that the March is about over, and kinda shocked that I've stayed to the end," said Doug. "I felt it was important to walk the final day in solidarity with you. I stayed out of a sense of obligation to the cause, to the group, and to you."

"Thanks, Doug, that means a lot to me," I respond, genuinely moved. "But despite the nuttiness, we're on the verge of accomplishing something pretty incredible, don't ya think?"

"Yeah, with all we've been up against, it's kind of amazing that we're here," Doug agrees.

We walk a short distance in silence, then Doug says, "I debated whether to tell you this, but I think I should." He pauses.

I look at Doug suspiciously. "Ok, spit it out."

"A couple nights ago there was an all-camp meeting," says Doug. "The rally at the White House came up, and some marchers wanted to interrupt your speech."

"Wow!" I say. "That's incredible. Climate activists heckling climate activists. No wonder we're losing to Big Oil," I say.

"Both Izzy and I spoke out against the idea," continued Doug. "Judging by the heads nodding, most marchers seemed to agree it'd be rude and counterproductive. Some of the newer marchers just looked bewildered. The mood at the meeting was miserable. I felt like I was watching the group enter its final death spiral."

I think about that as we march. Why so much discord? It's something I've pondered a lot this past week. How ironic that a community called One Earth Village became so fractured and divided, and that our precipitous descent into dysfunction grew worse and worse over time.

"Death spiral," I say, laughing a bit. "That sums it up pretty well, Doug. You know, the structure we set up at the start of the March was good, damn good. But it wasn't enforced, and that's my fault. I should have been a marcher *or* the director, not both. And it was insane for me to try to broadcast my talk show. I thought I could juggle all that, but I was wrong. It left little time to connect with other marchers and keep us focused."

"That might be true, but I'm not sure you or anyone else could have stopped the March from unraveling the way it did," observes Doug. "Everything on the March was more intense than in normal life."

For sure, relationships that would gel over the course of months in "normal life" formed in a few days in the raw, primal cauldron of One Earth Village. Some of the friends I made this year will last a lifetime. Yet bonds also disintegrated more quickly. I think of Jimmy and Kim, two of my best friends at the beginning of the March. Those friendships may be irreparably damaged. And Lala! It's hard to fathom how a relationship that started with such promise could have dissolved into such deep animosity.

I think about how lonely and isolated people are in modern America. The simple, intimate, communal life of the March packed a magic that's missing in most people's lives. Yet every assemblage of humans — modern or primitive — needs structure, rules, and an enforcement mechanism to prevent chaos.

"My experience on the March was a lot like my time with the Occupy Wall Street movement a few years back," Doug says. "I really liked the people, but things fell apart quickly. I watched people get so impassioned over their own issues that they tore the group apart from the inside."

"I remember Occupy, too, and you're spot on," I say. "Some marchers saw One Earth Village as a hippy commune. Honestly, we would have been more successful managed like an army battalion."

Doug laughs, and we joke about some of the young, more rebellious marchers standing at attention while John Abbe drops to the ground to do twenty push-ups for burning the oatmeal.

"We could have way too much fun with this," I laugh. "But seriously, that's the challenge in a free society, isn't it? How to allow the maximum amount of personal liberty without endangering the well-being of the community as a whole. An army battalion isn't the answer."

"Neither is a hippy commune or the consensus-style governance of Occupy," says Doug.

"In terms of the American experience, I think the best structure is a traditional town hall," I say. "That's what we tried to do with the March, but, well..." My voice trails off.

With just over a mile to go, we pass the statue of Mahatma Gandhi in front of the Indian Embassy. It depicts Gandhi on the 1930 Salt March, striding forward, walking stick in his right hand. The inscription reads, "My life is my message."

I smile. How appropriate that Gandhi should march with us for one stride, frozen in bronze and in time, on the final 1,500 steps of our March. I glance at my satchel, the gift from Gandhi's granddaughter, Sumitra Kulkarni. As with the March itself and each of us who've been with One Earth Village since Los Angeles, it's remarkable that this satchel has survived the journey.

After a few moments, I look away, embarrassed, pained that so often over these past eight months, my life has failed to be the message I want it to be for the world. My inability to hold the March community together. My frequent bouts of anger. My propensity to take on too much work — a common form of self-violence in the rat race of modern American life.

I've made so many mistakes. Yet I'd rather flaunt my mistakes than pretend they don't exist, rather confess them courageously in the full light of day than bury them in a closet, rather learn from them and hope not to repeat them in the future.

A MEMOIR FROM THE GREAT MARCH FOR CLIMATE ACTION

Perhaps my decision to leave the March in Hagerstown was also a mistake. But it felt right at the time, and it feels right now, mostly. Having six days and 85 miles to myself — where my footsteps became more of a pilgrimage than a march — helped bring the past eight months into perspective.

I understand love better, or at least I feel I've learned enough about love to get it right next time — or if not next time, then perhaps the time after that.

My appreciation for my home place has deepened. Iowa feels more special than ever, and I can't imagine ever wanting to be gone again for such a long span of time.

I realize that functional community requires striking that delicate balance between social intimacy and personal space, and that finding that sweet spot is an immense challenge constantly demanding adjustments.

Most important, I walk the final steps of the March confident that it's possible to accomplish what seems impossible. In the New Climate Era, where hope is desperately needed, the March showed that people can rise above all manner of physical, emotional, and political obstacles to do what must be done.

We enter the final blocks of the March. More people join as we turn south on 16th Street. A powerful wave of emotion crashes over me as the White House comes into view. I fight back tears. How is it that this diverse, dysfunctional, ridiculous crew of battered climate warriors just walked across America *and* arrived on schedule? It seems surreal, yet here we are, marching the last of seven million steps in what truly has been an epic journey.

When we arrive at Lafayette Square, I'm jerked back to reality. DC Action — the group in charge of preparations for the rally — has let us down badly. The stage is so small it barely holds two people. The microphone doesn't work. The young anarchist lets us use his megaphone, and I'm embarrassed that I spoke harshly to him earlier today.

About 250 people assemble for the rally. After music and a few remarks by Kathe, Kat Haber introduces me. I approach the podium, braced for the possibility of heckling or worse.

I begin by praising Mack, Jeffrey, Steve, and Miriam — the other four marchers who walked every step of the way. "For me, this commitment of unbroken footsteps was a necessary act of sacrifice. It demonstrated the deep commitment all people must embrace in the face of the climate crisis.

"But those of us who never missed a stride couldn't have done it without everyone else. We especially couldn't have done it without John and Mary, who made sure that camp came together every day, and without Sarah, Shari, Ki, our state coordinators, and the hundreds of volunteers along the way."

As I pan the crowd and pick out the marchers who've been with the March all or most of the way, even those who never or rarely walk, I see the hero in each of them. Despite all the conflict and hardship, I'm surrounded by an amazing group of individuals. These people chose to forsake the comforts of home life to make the strongest possible statement about the greatest existential threat humanity has ever faced. Yes, for all our shortcomings, this is truly a squadron of super heroes.

"We've had a lot of divisions within the March," I continue. "We've gotta move beyond that. We've gotta support each other. We're one movement. If you aren't down with direct action, you still need to support those who are. If you aren't down with political action, you still need to help those who lobby and run for office.

"The truth is, we're running out of time. The climate crisis grows more and more urgent with every passing day. Nonetheless, I'm optimistic, especially when I think of the thousands of people we touched from coast to coast. All across the country, there are hearts and minds we helped change, people we helped inspire.

"Remember Thomas Frank? Remember our action in front of the BP refinery south of Chicago? Thomas told me that some of the local people who marched with us that day were so moved they're now engaged and mobilizing others to speak out against BP. That's just one of many, many examples.

"We've made a huge difference, but our work's not done. We've got to brace ourselves for a long, grueling fight. Marchers, you've worked so hard and walked so far. Please, take a little time off —

but just a little. We've gotta keep pushing. We've gotta continue to deliver the message that we're gonna fight until we win climate action and climate justice, because our survival — and the survival of this beautiful web of life we call home — depends on it."

As I leave the stage, I breathe a sigh of relief that I wasn't heckled — relieved mostly because it would have embarrassed the local people who came to support us.

Next, Kathe invites marchers to the stage to read index cards written by people we met along the way, expressing their concerns about climate change. After that, marchers form a circle and sing "This Pretty Planet." It's a beautiful song, but other than marchers, almost no one knows it. Some marchers begin to dance, and it's not clear whether the rally is over or if there's more to come. Things devolve into people milling about — visiting, crying, hugging. Symbolically, the confusing, chaotic conclusion of the rally feels like the perfect ending for the March itself.

I chat with some of the people who came to support us, then wander toward the White House for a few photos. The March is done and I'm spent. Some marchers will remain in DC for a week of direct action against FERC. I'll leave tomorrow and take two weeks to rest and recover. Then I'll join other Iowans fighting the Dakota Access pipeline.

As I walk around the park, not quite sure what to do next, I wonder how this experience has changed other marchers. For all its shortcomings and heartaches, the incredible commitment and sacrifice of those who marched was, without a doubt, the most focused expression of power and love I've ever been part of.

As Kat later wrote, "We were such gypsies, anomalous and driven through the dreamlike state of national complacency. Urgently aware were we, we who are vast, deep, playful, singing, paying attention right here, right now, aware of our inner connectedness. We gave our footsteps all our intention to nudge America awake. Peace our method, we were connective conversations and songs for those who slowed down enough to know what was absolutely important. What a gift it was to wander with the gypsies for a year!"

Maybe some marchers will remain climate-action gypsies, traveling from place to place, plugging in where needed. Maybe others will settle back in their communities and build support for change at the local level. Maybe some will run for office. I'm certain many will risk arrest, perhaps repeatedly. It's all good. It's all needed. Time is short, and full comprehension of the precarious urgency of climate change hasn't sunk in for most Americans. It's up to us — those who grasp the gravity of the moment — to goad others, politely yet persistently, to take action.

My deepest hope for my community of marchers — my bold, loving, crazy, incorrigible, and passionate family of battered climate warriors — is that we find happiness wherever we land and however we work for a better world, and that each day we remember that we're all cut from the same cloth, rolled from the same dust, children of the same bloodied cross or flowering lotus tree on a planet, our home, that we desperately need to love and protect.

INDEX

A Game of Thrones, George R. R. Martin, 101
A Sand County Almanac, Aldo Leopold, 133
A Walk in the Woods, Bill Bryson, 265
Abbe, John, 30, 46, 96, 248, 274, 275, 326
Abbot, Karen, 99, 100
ABC News, 273
Above All Else (movie), 179
Adair County, Iowa, 191
Addams, Mary, 45, 46
Albuquerque, New Mexico, 59, 95, 125, 126, 128
"All of Me," Billie Holiday, 137
All People's Christian Church, Los Angeles, California, 31
Allegheny Mountains, 278, 293
Alliquippa, Pennsylvania, 280
Amana Colonies (Amana), Iowa, 210, 211
American Legion, 278
Amish, 63, 258, 259, 260
Anamosa, Iowa, 209

Appalachian Mountains (Appalachia), 279, 289, 318
Aral Sea, 262
Arizona Department of Human Services, 81
Arizona Department of Transportation (Arizona DOT, or DOT), 77, 78, 107
Arsenal Bridge, Davenport, Iowa, 214
Atlantic, Iowa, 191
Atlantic Ocean (the Atlantic), 23, 136, 183
Bach, Richard, 323
Barber, Mary, 159
Barley and Rye Bistro, Moline, Illinois, 215
Bartley, Nebraska, 175
Beaumont, California, 64
becomenomad.com, 305
Bedouin, 181
Belgium, 182, 184
Ben and Jerry's, 203
Ben Avon, Pennsylvania, 280
Berry, Wendell, 258

Bethesda, Maryland, 313, 322, 323
Betts, Jimmy, 9, 31, 40, 44, 46, 54, 66, 74, 75, 76, 81, 118, 124, 129, 130, 131, 158, 159, 161, 162, 173, 177, 189, 236, 240, 266, 269, 325
Bible, 94
Biden, Joe, 6, 7, 297
Black (Blacks), 135, 230
Bob Kerrey Bridge, Omaha, Nebraska, 186
Boonsboro, Maryland, 309
Booth, Leonardo, 61, 141
Booth, Trish, 61, 141
Boy Scouts of America (Boy Scouts), 92
Boyko, Mark, 99, 100
Breezewood, Pennsylvania, 301, 304, 314
Bridgewater, Iowa, 191
British Petroleum (BP), 229, 230, 247, 248, 264, 328
Brody, Al, 145, 155
Bronx, New York, 122
Brooklyn, New York, 126
Brookner, Anita, 312
Brown, Edwin, 314
Brown, Greg, 303
Browne, Jackson, 281
Brussels, Belgium, 183
Bryson, Bill, 265
Buddhist, 144, 145, 167
Bureau of Land Management (BLM), 24, 69, 70, 72, 74, 78
Burn Zone, Danny Lyon, 125
Burton, Jonathan, 314

Bushwick, Ben, 46, 49, 59, 128, 243
Butler, Pennsylvania, 252, 278
C&O Canal Trail, Maryland, 4, 254, 310, 312
Cabazon, California, 39, 64
Cafe Dodici, Washington, Iowa, 167
Cahokia Mounds, Mississippi, 107
Cairo, Egypt, 19
Camino del Norte a Chimayo, New Mexico, 139
Camp Fire Long Beach, Long Beach, California, 23, 25, 26, 28
Camp Shadow Pines, Heber, Arizona, 95, 96
Canada (Canadian), 8, 170
Canterbury Tales, Geoffrey Chaucer, 181
Canyon de Chelly, Arizona, 107
Carter, Majora, 227
Castro, Fidel, 284
Catholic Church (Catholicism, Catholic), xv, 18, 101, 108, 139, 145, 167, 259
Cazares, Fernando, 228, 229, 249, 253, 272, 275, 282, 291, 292, 293, 294, 295, 313, 323
Cedar River, Iowa, 212
Center for Applied Conflict Management, Kent State University, 275
Chantel, Annamarie, 277
Chaucer, Geoffrey, 181, 182
Cheshire, Brandon, 46
Cheshire, Erica, 46, 77, 95
Cheyenne, Wyoming, 18

Chicago, Illinois, 219, 220, 221, 224, 225, 226, 227, 228, 231, 247, 248, 264, 266, 286, 312, 328
Chicago Board of Trade, 227
Chicago Botanical Center, Chicago, Illinois, 225
Chief Motel, McCook, Nebraska, 172, 173
Chimayo, New Mexico, 140, 317
China, 171
Chopin, Frederic, 20, 36, 136, 205, 206, 315
Chopin Nocturne Opus 27 #1 (Chopin's "The Great Nocturne"), 205, 206, 207
Chopin Nocturne Opus 9 #2, 36
Christianity (Christian), 172, 182, 301
Church, Ted, 209
Church, Troy, 203, 205, 209
Church of the Redeemer, Pittsburgh, Pennsylvania, 285
Citizens' Climate Lobby, 225
Claremont, California, 36, 89
Clay, Peter, 43
Cleghorn, Stephen, 281
Cleveland, Ohio, 267, 269, 277, 318
Cleveland Heights, Ohio, 272, 273, 274
Climate Justice Gypsy Band, 46, 126, 212, 224, 236, 240
Clinton, Hillary, 23
Coal Miners Cafe, Jennerstown, Pennsylvania, 297, 298

Cobbett, William, 166
Coelho, Paulo, 62, 247
Cohen, Leonard, 284
Cold Turkey (movie), 191
Colorado River, 46, 74, 76
Colorado River Indian Tribes, Parker, Arizona, 47, 77
Colorado Springs, Colorado, 145, 159, 236
Communism (Communist), 146, 262, 263
Concho, Arizona, 99
Cook, Bob, 32, 46, 67, 116, 117, 234
Cooley, Doug, 252, 307, 324, 325
Cornell College, Mount Vernon, Iowa, 187
Coulson, Ki, 23, 24, 30, 66, 161, 177, 328
Cox, Nigel, 205
Creekwater, Mark, 50, 168, 169
Crow, Valerie, 267
Culbertson, Nebraska, 169, 172, 239
Cumberland, Iowa, 191
Czerwiec, Jeffrey, 25, 40, 46, 152, 243, 244, 251, 265, 281, 282, 284, 321, 328
Daas, Ram, 143
Dagmar Hotel, Hagerstown, Maryland, 307, 308
Dakota Access Pipeline, 179, 201, 329
Daley Plaza, Chicago, Illinois, 227
Damon, Matt, 277
Darion, Joe, 17

Darlington, Pennsylvania, 278
Davenport, Iowa, 209, 213, 214, 215, 246
Davis, Luke, 39, 46, 236
Davis, Marie, 25, 46, 76, 154, 178, 222
DC Action, 289, 327
DeCamp, Mary, 116, 129, 296, 328
Delaware County, Oklahoma, 176
Democratic Party (Democrat), 171, 172
Denver, Colorado, 139, 159
Des Moines, Iowa, 5, 7, 9, 11, 12, 13, 20, 30, 31, 63, 81, 104, 105, 113, 115, 124, 143, 164, 176, 179, 185, 186, 187, 191, 192, 197, 199, 200, 203, 204, 205, 207, 208, 209, 216, 226, 261, 286, 290, 301, 307
Detroit, Michigan, 261, 262, 263, 265
Devil's Tower, Bear Lodge Mountains, Black Hills, South Dakota, 107
Diggy, 203, 204, 207, 245
Dimi Espresso Coffeehouse, Payson, Arizona, 93
Diné, 142
Don Quixote, 64
Donner Party, 8
Dowa Yalanne, New Mexico, 108
Downton Abby (television series), 12
Downtown Owl, Chuck Klosterman, 219
Dublin, Ireland, 297

Dulcinea, 17, 86, 316, 317
Duncan, David James, 23
Durant, Iowa, 213
Duthoo, Annika, 182, 183, 184
Early, Jubal A., 314
East Chicago, Chicago, Illinois, 229, 231
Eco Commode, 24, 67, 100, 113, 114, 117, 143, 173, 191
Egypt, 176
Elkhart County, Indiana, 258
Elm Street Urban Park, Bethesda, Maryland, 323
Elmore, Ohio, 267, 268
Energy Barn, York, Nebraska, 179
Energy Transfer Partners, 179
England (English), 70, 84, 170, 181, 308
English Channel, 183
Environmental Protection Agency (EPA), 159
Epicurus, 315
Erickson, Kelsey, 46, 80, 81, 82, 106, 117, 129, 139, 158, 161, 162, 163, 173, 175, 177, 227, 243, 253, 259, 266
Erie, Illinois, 219
Evans, Kenny, 92
Fallon, Ben, 202, 203, 204, 207, 208, 245
Fallon, Ed (Fallon), 39, 42, 44, 46, 50, 53, 54, 55, 56, 57, 59, 60, 61, 74, 80, 81, 106, 129, 130, 136, 162, 163, 168, 177, 189, 190, 193, 199, 209, 222, 234, 235, 236, 243, 245,

249, 250, 266, 267, 273, 274
Farmer Miller, 133, 134
Fascism, 122
Father Comaine, xv
Federal Energy Regulatory Commission (FERC), 274, 275, 301, 329
Field Museum, Chicago, Illinois, 225
Fionna, 84
Fitzgerald, F. Scott, 3
Flanders (Flemish), Belgium, 182
Foley, Kim, 31, 42, 46, 53, 68, 75, 81, 82, 100, 160, 177, 189, 191, 225, 313, 325
Forrest Gump (movie), 155
Fort Carson, El Paso County, Colorado, 159
Fort Garland, Colorado, 147, 148, 149, 150, 151, 235
Fort Morgan, Colorado, 163
Franciscan, 110
Frank, Thomas, 229, 230, 328
French Alps, 167
Front Range, Colorado, 158, 160
Fuller's Cafe, McCook, Nebraska, 173
Furst, Bobby, 66, 67
Gandhi, M.K., xiii, 14, 53, 112, 123, 271, 276, 326
Gardiner, Bob, 105, 106
Gary, Indiana, 229, 230, 231
General Electric, 146
George Washington Bridge, New York, 262
"Georgia on My Mind," Ray Charles, 137
Germany, 210
Gibbon, Nebraska, 176
Glenn, Sean, 30, 46, 47, 50, 127, 215, 229, 230, 231, 236, 250, 261, 262, 264, 266, 277
Gluba, Bill, 213, 214, 215, 246
Goji Eco Lodge, New Mexico, 143
Goldman, Charles, 96, 97, 128, 185, 186, 213, 270
Good Shepherd Catholic Church, Toledo, 267
Grace, 5, 11, 16, 22, 37, 68, 73, 85, 86, 104, 105, 119, 144, 180, 185, 186, 190, 192, 195, 196, 197, 198, 199, 207, 208, 216, 218, 279, 285, 286, 287, 317
Graham, Lindsey, 29
Grange County, Indiana, 258
Grants, New Mexico, 117, 118, 119
Gray's Lake, Des Moines, Iowa, 9
Great Hunger of 1845 (the famine), 84, 135
Great March for Climate Action (Climate March), vii, xi, xiii, xiv, xvi, 3, 8, 14, 31, 69, 77, 128, 153, 170, 177, 183, 214, 228, 229, 246, 248, 263
Great Peace March (Peace March), xiii, 8, 34, 35, 176
Greek mythology, 18, 125

Greene, Monica, 309, 310, 311, 312
Greene, Roy, 309, 310
Greenfield, Iowa, 186, 191, 192, 195
Greensburg, Pennsylvania, 295
Grinnell, Iowa, 196, 207, 208
Gulf of Mexico, 164, 214, 262
Haber, Kat, 61, 236, 327, 329
Hach, Rob, 115
Hagerstown, Maryland, 305, 307, 309, 324, 327
Hancock, Landon, 275, 276
Hanuman Temple, Taos, New Mexico, 142, 143
Harkin, Tom, 149
Heber, Arizona, 96
Heffernen, Zach, 80
Henderson, Colorado, 160
"Hey, That's No Way To Say Goodbye," Leonard Cohen, 284
Highway 160, Colorado, 151
Highway 51, Indiana, 267
Highway 62, California, 40, 65, 66, 72
Highway 87, Arizona, 92
Hindu, 145
Hippie, Martin, 34, 66, 69
Hispanic, 30, 148, 230
Hobbits, xvii
Hough Solomon, Zach, 243
Howard, Pablo, 46, 113, 114
Hrdina, Shari, 8, 40, 63, 82, 99, 130, 178, 207, 328
Hunn, Allison, 309, 319
Huxley, Aldous, 143

Ildan, Mehmet Murat, 91
Illusions: The Adventures of a Reluctant Messiah, Richard Bach, 323
Imperial Steel, 228
Impey, Maureen, 183
India (Indian), 276, 308
Indian Embassy, 326
Indigenous (Native American, Native, "Indian"), 29, 47, 58, 109, 110, 111, 112, 113, 126, 135, 142, 145, 147, 148, 193, 194, 267, 292, 293, 294
Interstate 10, California, 64
Interstate 70, Pennsylvania, 302
Interstate Highway System (Interstate), 12, 13, 14, 64, 158, 162
Iowa City, Iowa, 211, 212
Iowa Legislature (the Legislature), 210
Iowa Policy Project, 187
Iowa State Capitol (Iowa Statehouse), 70, 71, 188, 197, 199, 200, 203, 207, 244, 245
Iowa State University, 6, 211
Ireland (Irish), xvi, xvii, 74, 84, 105, 113, 122, 134, 135, 147, 181, 183, 272, 292
Issa, Kobayashi, 120
Italian, 25, 167
Jackson, Jesse, 191
Jacogshagen, Keith, 156
Jagger, Mick, 195
James, Debaura, 25, 46

Japan, 171, 212
Jefferson, Thomas, 304
Jerusalem, 220
Jesus, 172, 220, 300, 301
Jody, 298, 299
Joel, Billy, 84
Jorgensen, John (John J), 46, 47, 49, 53, 70, 72, 77, 105, 106, 129, 162, 177, 217, 236, 259, 260, 328
Joshua Tree, California, 40
Joshua Tree National Park, California, 65
Joshua Tree Retreat Center, California, 66
Juliana, Kelsey, 243
June's Cafe, Heber, Arizona, 96, 97
Kagyu Milo Guru Stupa, New Mexico, 144
Kansas City Royals, 315
Karma (karmic), 21, 287
Karrer Park, McCook, Nebraska, 173
Kashia, Miriam, 46, 65, 82, 88, 161, 194, 211, 212, 221, 248, 250, 251, 265, 269, 275, 281, 282, 284, 289, 315, 321, 328
Kate, 102
Kendall, Jane, 45, 46, 91, 176
Kent, Ohio, 275
Kent State University, Kent, Ohio, 275, 276
Keystone XL pipeline, 9, 165, 175, 179, 24
Klosterman, Chuck, 219

Knuth, Robert, 193, 194, 196
Kristin, 19, 20, 21, 85, 120, 135, 144, 202, 203, 320
Krizek, Corky, 169, 170, 172, 173
Kubler-Ross, Elizabeth, 143
Kulkarni, Sumitra, 14, 53, 271, 326
Kuralt, Charles, 12, 116
L'Arche, France, 181
La Bajada, New Mexico, 133, 134
LA County Health Department, Los Angeles, California, 64
Lac-Mégantic, Quebec, 174
Ladora, Iowa, 209
Lafayette Square, Washington, DC, 257, 289, 307, 327
Lafferty, Liz, 46, 160, 177, 189, 225, 240, 313
Lagomarcino, Beth, 215, 216
Lagomarcino's, Moline, Illinois, 215
Laguna Pueblo, New Mexico, 125, 300
Lake County, Indiana, 230
Lake Erie, 262
Lake Michigan, 230
Lakota, 193
Lama Foundation, New Mexico, 144
Lawrence, D.H., 143
LeClaire Park, Davenport, Iowa, 213, 214
Lee, Jonathan, 38
Leesburg, Virginia, 255, 312, 313, 315
Leopold, Aldo, 133

Lewis, Iowa, 190, 191
Lincoln, Nebraska, 183, 186
Loess Hills, Iowa, 187
London, England, 77, 183
London, Jack, 318
"Lonesome Sundown," Tom Petty, 289
Los Angeles, California (LA), 8, 19, 14, 15, 23, 24, 30, 33, 36, 39, 76, 90, 92, 127, 149, 153, 169, 193, 194, 198, 200, 204, 209, 222, 226, 268, 290, 319, 324, 326
Lown, Loren, 164
Lujan, Lawrence, 142
Lyndon, Illinois, 221
Lynn, 21
Lynn Boys Club, Lynn, Massachusetts, 292
Lyon, Danny, 59, 125, 126, 127, 129, 131
Lyon, Nancy, 59, 126
Macedonia, Ohio, 274
Madison County, Iowa, 194
Mahkee, Wells, 108, 109
Malevich, Kazimir, 273
Mandela, Nelson, 294
Manhattan, New York, 262, 263, 264
March Council (Council), 34, 116, 129, 130, 187, 209
Marcontel, Bryant, 261
Marlboro College, Green Mountains, Vermont, 102
Martin, George R. R., 101
Martin, Steve, 45, 48, 61, 74, 75, 77, 86, 87, 88, 89, 91, 94, 95, 100, 116, 122, 123, 124, 130, 131, 132, 134, 140, 141, 155, 156, 157, 159, 160, 162, 163, 164, 165, 166, 168, 169, 173, 194, 234, 238, 262, 265, 281, 292, 313, 321, 322, 323, 328
"Mary Had a Little Lamb," 223
Mason & Hamlin (piano), 205
McCook, Nebraska, 169, 170, 172, 174
McCook Gazette, 174
McDaniel, Anita, 148, 149
McGrath, Shaun, 159
McGuire, Dana, 225, 243, 296, 301
McKean, Andy, 209
McKibben, Bill, 6, 7, 211
Meckley, Faith, 142, 236, 243, 247, 249, 267, 268, 281
Mediterranean, 34, 99, 222
Mediterranean Sea, 183
Methodist, 145, 228
Mexico (Mexican), 8, 126, 153, 171, 228, 293, 296
Middle Earth, xvii
Middlebury, Indiana, 258
Middletown, Maryland, 310
Midwest (Midwestern), 27, 31, 160, 191, 315
Mika, 79, 80, 203, 205
Miller, Taylor, 66
Mississippi River (Great River), xvi, 213, 214, 217, 218, 219, 226, 246
Missouri River, xvi, 186, 187, 205, 241

Mitchellville, Iowa, 207
Mitzner, Johnny, 176
Moeller, Paul, 247
Moffat, Joe, 135
Moffat, Kathleen, 135
Mogelgaard, Izzy, 46, 105, 106, 177, 226, 240, 243, 325
Mogollon Rim (the Rim), Arizona, 90, 93, 94, 158
Mojave Desert (the Mojave), 24, 34, 44, 45, 46, 63, 66, 67, 69, 70, 71, 72, 74, 76, 77, 115, 149, 158, 162
Moline, Illinois, 215
Momaday, N. Scott, 107
Montpelier, Ohio, 265, 266
Moore, George A., 209
Mormon, 95
Moscow, Iowa, 212, 213
Mount Everest, 94
Mountain Watershed Association, 278
Mudheads, 109, 110
"My Life," Billy Joel, 84
National Famine Museum, Strokestown, Ireland, 84
Navajo, 88, 196
Nayowith, Bruce, 305
Necker, Chuck, 89
Necker, Linda, 89
Nelson, Derek, 113, 114
New Age, 143, 145
New Buffalo Center, Arroyo Hondo, New Mexico, 61, 143, 234
New Climate Era, xiv, xvii, 88, 89, 111, 112, 194, 327

New Melleray Abbey, Peosta, Iowa, 121
New York, New York (New York City), 77, 91, 126, 162, 238, 260, 261, 262, 265, 266, 269
New York (state), 142
Newago, Mike, 110
Newago, Sticker, 110
Newton, Iowa, 209
Nixon, Richard, 76
North Olmstead, Ohio, 269, 270
Novak, Stephanie, 278
Oakland, Iowa, 189
Obama, Barack (the President, the Obamas), 201, 259, 297, 309, 310, 319
Oberlin, Ohio, 269
Occupy Wall Street (Occupy), 325, 326
Odysseus, 18
Oedipus complex, 305
Ogallala Aquifer, Nebraska, 179
Ohio National Guard, 276
Ohio River Road, Pittsburgh, Pennsylvania, 281, 282
Ojibwe, 18, 110, 111
Omaha, Nebraska, 186
One Day in the life of Ivan Denisovich, Alexandr Solzhenitsyn, 95
One Earth Village, 26, 27, 66, 72, 87, 90, 111, 112, 113, 115, 125, 130, 177, 194, 212, 219, 224, 323, 325, 326
Osterberg, David, 187, 188, 189, 190, 191, 192, 193,

199, 200, 242, 243
Pacific Ocean, 30, 191, 200
Packer, Mel, 281
Palazzolo, Lala, 25, 27, 33, 34, 35, 36, 46, 50, 61, 63, 68, 73, 76, 79, 81, 99, 117, 118, 119, 124, 130, 154, 162, 173, 177, 188, 189, 190, 191, 192, 193, 195, 196, 200, 207, 221, 222, 231, 236, 243, 266, 267, 268, 269, 289, 290, 291, 292, 294, 306, 325
Paris, France, 77
Parker, Arizona, 46, 47, 74, 75, 77, 79
Parker, Dorothy, 33
Parker, Jordan, 219, 221, 224, 225
Patton, George S., 146, 147
Pauley, Dennis, 215
Payson, Arizona, 51, 89, 90, 91, 92, 93, 94, 106, 280, 300
Payson Christian School, Payson, Arizona, 92
Payson Roundup, 92
Peace Corps, 187
Pearl Harbor, 212
People's Climate March, 163, 260, 262, 266
Peterson, Clifford, 260
Peterson, Lisa, 260
Petty, Tom, 289
Phillips, Ethan, 39, 46, 57
Phoenix, Arizona, 79, 86, 88, 89, 91, 99, 194
Pietist, 210
Pilgrims Way, England, 181
Pisano, Tony, 61, 86, 117, 134, 140, 142
Pittsburgh, Pennsylvania, 267, 280, 281, 282, 285, 295, 299
Pittsburgh Penguins, 295
Point of Rocks, Maryland, 310, 311, 312
Port of Los Angeles, Los Angeles, California, 23
Post-Victorian Man (P-V Man), 17, 18
Potawatomi, 228, 293
Potomac River (the Potomac), 310, 312, 313, 314, 319, 320
Prairie View High School, Henderson, Colorado, 160
Promised Land (movie), 277
Quayle, William A., 175
Queens, New York, 126
Rachel, 315, 316
Redlands, California, 62
Redlands United Church of Christ, Redlands, California, 63
Remembering, Wendell Berry, 258
Republican Party, 171, 172, 178, 209
Revere, Paul, 8
Rhodes, Silvia, 275
Ricky, 308
Rin-Tin-Tin, 137
Riordan, Rick, 151
Ritual Cafe, Des Moines, Iowa, 203, 244
Rocheleau, Ahni, 61, 95
Rock Island, Illinois, 215
Rock Island Botanical Center, 215
Rocky Mountains, "Rockies", 8, 15, 132, 149

Roggin, Colorado, 162
Route 66, 36, 57, 120
Russian (Russians), 146, 148, 183
Rye, Arizona, 92
Sage Granada Park United Methodist Church, Alhambra, California, 33
Salt March, India, 326
Sandahl, Karin, 27, 185
Sangre de Cristo Mountains, 145, 147, 151, 154
Santa Fe, New Mexico, 95, 130, 133, 134, 136
Santa Fe National Forest, New Mexico, 141
Santa Fe River, New Mexico, 133
Santuario de Chimayo, Chimayo, New Mexico, 60, 139, 140
Saunders, Lacy, 136, 137
Scandinavia, 182
Scattergood Friends School, West Branch, Iowa, 212
Schenley Bridge, Pittsburgh, Pennsylvania, 285
Schenley Park, Pittsburgh, Pennsylvania, 284
Scottsdale, Arizona, 300
Scovill Avenue Neighborhood, Cleveland, Ohio, 273, 274
Seidl, Frank, 67, 68
Seidl, Laurel, 67, 68
Seneca Lake, New York, 142
Serbia, 285
Shipshewana, Indiana, 259
Shipshewana Auction and Flea Market, Shipshewana, Indiana, 259
Shoshone, 18
Sierra Club, 225
Simpson, Bob, 247
Sinai Desert, Egypt, 181
Sirens, 18
Sister Antoinette, 101
Snowflake, Arizona, 95, 98
Snyder, Ken, 46, 82, 113
SoCal Climate Coalition, 30
Solzhenitsyn, Alexandr, 95
Sonoran Desert, 49, 77, 93, 115, 129, 219
South Bend, Indiana, 231
Southern Methodist University Retreat Center, New Mexico, 141
Sowden, Phil, 183
Spain, Sarah, 4, 5, 46, 47, 51, 54, 58, 89, 90, 98, 99, 107, 112, 113, 114, 115, 130, 133, 142, 143, 151, 162, 166, 169, 173, 192, 195, 196, 199, 228, 259, 260, 266, 267, 269, 272, 273, 274, 299, 300, 301, 313, 321, 328
Spanish language (Spanish), 203
Spatti, Ray, 91, 92
St. Johns, Arizona, 103, 105, 108
Stalin, Josef, 12
Steinbeck, John, 70
Stewart, Lee, 46, 164, 243, 265, 266
"Summertime," George Gershwin, 137
Supremist Composition, Kazimir Malevich, 273

Sustainable Fort Carson, Fort Carson, Colorado, 159
Taize, France, 181
Taos, New Mexico, 95, 117, 142, 143, 144
Taos Pueblo, New Mexico, 142
Tchaikovsky, Pyotr Ilyich (Tchaikovsky), 147
The Alchemist, Paulo Coelho, 62
The Brook Kerith, George Moore, 209
The Changing Prairie, Anthony Joern (editor), 156
The Corner (café), Greenfield, Iowa, 192
The Glass Outhouse Art Gallery, Twentynine Palms, California, 43, 44, 67, 77
The Great Gatsby, F. Scott Fitzgerald, 3
The Prairie and the Sea, William A. Quayle, 175
The Throne of Fire, Rick Riordan, 151
Third U.S. National Climate Assessment for the Southwest, 88
"This Land Is Your Land," Woody Guthrie, 76
"This Pretty Planet," Tom Chapin, 329
Thompson, Kathe, 53, 61, 209, 223, 248, 272, 306, 327, 329
Thoreson, David, 27
Tilly, 243, 296
To Build a Fire, Jack London, 318

Todd, Judy, 59, 128, 131
"To Each His Dulcinea," *Man of La Mancha,* 17
Toledo, Ohio, 249, 266, 267, 269
Tolkien, J.R.R., xvii
Tolstoy, Leo, 284
Tompkins, Berenice, 57, 236, 239, 243, 290
TransCanada, 9
Treynor, Iowa, 187, 188
Truchas, New Mexico, 61, 140
Trump, Donald, 27
Tucson, Arizona, 70
Twentynine Palms, California, 66, 67
U.S. Department of Defense (Department of Defense), 13, 146, 148
U'Pritchard, Gavain, 115, 219, 220, 221, 225, 227, 232, 243, 262, 266, 294, 296, 301, 305, 306
United Nations (UN), 292
US Army (Army), 147, 148, 149, 159, 189, 310
US Coast Guard, 314
USS Constitution, 146
USSR, 262
Utes (Ute people), 147, 148, 149
Valero, 29, 30
Ververis, Chris, 159, 240, 243
Victorian Man (Victorian, Victorian Era), 17, 18, 21, 86, 316
Vidal Junction, California, 74
Vietnam War (Vietnam), 146, 147, 276

Vilsack, Tom, 70, 71, 74, 78, 309
Vom Dorp, Elizabeth, 143
Vom Dorp, Eric, 143
Walcott, Iowa, 213
Wallace, Henry, 191
Walmart, 111, 211
Warner, Bob, 309
Warren, Kelcy, 179
Washington, Iowa, 167
Washington, DC (DC), 3, 8, 30, 97, 115, 116, 142, 153, 162, 179, 193, 226, 228, 256, 258, 259, 269, 274, 289, 290, 301, 305, 309, 310, 312, 319, 324, 329
West Branch, Iowa, 212
Western Trails Ranch Restaurant, Morristown, Arizona, 79
White (Whites, Anglo America), 135, 136, 147, 148
White House, 3, 4, 152, 221, 257, 258, 259, 289, 290, 302, 307, 309, 310, 312, 319, 323, 324, 327, 329
White Mountains, 98
White's Ferry, 255, 313, 314, 319
Whiting, Indiana, 229, 247
Wiggins, Colorado, 162, 190
Wilkins, Mack, 46, 47, 50, 64, 65, 68, 73, 92, 128, 215, 229, 230, 231, 236, 240, 243, 250, 265, 266, 267, 272, 277, 281, 282, 284, 321, 328
Wilmington Waterfront Park (Waterfront Park) Los Angeles, California, 29, 38
Wilton, Iowa, 213
Windwalker, Mary, 98, 99
Windwalker, Tim, 98, 99
Winterset, Iowa, 192, 193, 195
Wishart, Anna, 165, 166
Witt, Bill, 13
Wohlberg, Shira, 39, 46, 48, 47, 84, 61, 72, 73, 79, 82, 85, 86, 87, 89, 94, 95, 100, 106, 108, 116, 117, 118, 130, 131, 132, 134, 140, 165, 262, 263, 279
Woman Stands Shining, 142
World Series, 315
World War II, 122
York, Nebraska, 241
Youngstown, Ohio, 277
"Your Town Now," Greg Brown, 303
Zahrt, David, 44, 46, 177
Zambrano, Michael, 46, 106, 129, 130, 302
Zuni (Zuni Pueblo), New Mexico, 54, 103, 107, 108, 110, 111
Zuni Correctional Facility for Juveniles, Zuni, New Mexico, 108, 109, 110

CPSIA information can be obtained
at www.ICGtesting.com
Printed in the USA
BVHW092352171118
533042BV00001B/1/P